Peter Berger on Modernization and Modernity

With particular attention to his work on modernization and modernity as construed by a sociologist of knowledge, this book offers a sympathetic exposition and evaluation of Peter Berger's work as one of the world's most accomplished and influential sociologists.

In the context of an examination of Berger's ongoing work on the social construction of reality, styles of consciousness, the role of science-based technology, pluralism, and other pertinent topics, the author also considers Berger's unique and thoughtful approach to research and theorizing. Berger's method of "sociological tourism," which departs sharply from the current emphasis in the social sciences on ever more complex and ostensibly rigorous statistical procedures, provides a refreshing move away from the increasingly esoteric and sometimes alienating methodological self-consciousness that characterizes contemporary sociology.

With this distinctive approach, this book will appeal to scholars and students of sociology who share Berger's interest. The importance of modernization and modernity on a world scale is undeniable, and a deeper understanding of their nature and consequences, will also benefit members of the intelligent laity who are not sociological specialists but are open to new ideas that are clearly explained.

Robert Bickel is Professor Emeritus of Advanced Educational Studies at Marshall University, Huntington, West Virginia, and author of *Multilevel Analysis for Applied Research* and *Classical Social Theory in Use.*

Routledge Studies in Social and Political Thought

For a full list of titles in this series, please visit www.routledge.com/series/RSSPT

119 Theories of the Stranger
Debates on Cosmopolitanism, Identity and Cross-Cultural Encounters
Vince Marotta

120 Deciphering Goffman
The Structure of his Sociological Theory Revisited
Ramón Vargas Maseda

121 Planet Utopia
Utopia, Dystopia, and Globalisation
Mark Featherstone

122 Weber, Schumpeter and Modern Capitalism
Towards a General Theory
John Love

123 The Cultural Contradictions of Anti-Capitalism
The Liberal Spirit and the Making of Western Radicalism
Daniel Fletcher

124 Peter Berger on Modernization and Modernity
An Unvarnished Overview
Robert Bickel

125 Imaginaries of Modernity
Politics, Cultures, Tensions
John Rundell

126 Complexity, Society and Social Transactions
Developing a Comprehensive Social Theory
Thomas Whalen

Peter Berger on Modernization and Modernity

An Unvarnished Overview

Robert Bickel

Professor Emeritus
Advanced Educational Studies
College of Education
Marshall University

Routledge
Taylor & Francis Group
LONDON AND NEW YORK

First published 2018 by Routledge

2 Park Square, Milton Park, Abingdon, Oxfordshire OX14 4RN
52 Vanderbilt Avenue, New York, NY 10017

Routledge is an imprint of the Taylor & Francis Group, an informa business

First issued in paperback 2018

British Library Cataloguing in Publication Data
A catalogue record for this book is available from the British Library

Library of Congress Cataloging in Publication Data
A catalog record for this title has been requested

ISBN: 978-1-138-08520-6 (hbk)
ISBN: 978-0-367-20816-5 (pbk)

Typeset in Times New Roman
by Taylor & Francis Books

Dedicated, with love, to my wife Lisa Higgins.

Contents

Foreword

Peter Berger has long been one of the world's most influential and accomplished sociologists not only in the United States but much of the rest of the world. He is one of an intellectually accomplished group who works in the sub-discipline known as the sociology of knowledge. Interested in the reciprocal, mutually implicative relationship between consciousness and social circumstances he is deeply immersed in the sociology of modernization and modernity and the sociology of religion (Berger [Ed.], 1990). Berger's work on modernization and modernity, done with the theoretical, substantive, and, in his own way, methodological wherewithal of a brilliant sociologist, provides the material for this evaluative overview.

I think that anyone interested in our complex, conflict-ridden world, whether scholar, teacher, advocate, or intelligent member of the laity, would benefit from reading Berger's prolific treatment of modernization and modernity. If *Peter Berger on Modernization and Modernity* succeeds in approximating justice to Berger's contribution it will have accomplished what I had hoped to do. If it encourages the reader to turn to Berger's original work, that is all to the good.

I think that one of the reasons I find Berger's work fascinating, is because he is outside the sociological mainstream. His preferred method, which he informally characterizes as "sociological tourism," departs sharply from the current emphasis in the social sciences on ever more complex and ostensibly rigorous statistical procedures. Nevertheless, I am convinced that Berger's work is far more influential, in a taken-for-granted way, than many sociologists appreciate. With remarkable books such as *The Social Construction of Reality* (Berger & Luckmann, 1966) Berger has substantially enriched the foundation of sociological understanding and practice.

1 Background: Berger's prose, presentation, and purpose

Peter Berger's contribution to contemporary sociology is truly substantial, but perhaps easy to overlook. His notion of the social construction of reality quietly pervades much of the material that today's sociologists produce (Berger & Luckmann, 1966). Much like Durkheim (1912/1995) Berger has tacitly contributed to shaping the nature of sociological discourse in a taken-for-granted way and to a degree that are not commonly acknowledged. He has done this in a workmanlike fashion, often in the company of like-minded colleagues. (See, for example, Berger & Kellner, 1981; Berger, Neuhaus & Novak, 1994; Berger & Huntington, 2002.) All the while, Berger's work has remained original, interesting, and more accessible than one expects from an author who addresses undeniably difficult issues.

Berger's prose style is neither elegant nor convoluted. Though occasionally a bit cluttered, for the most part it is a model of plain-spoken clarity. One must remember, however, that Berger is an intellectual and a professional sociologist, typically writing for those who fit into the same or similar categories. Moreover, from time to time and source to source he may resort to the use of gratuitous neologisms, something that, in my experience, often leaves an author and his readers poorly served.

Nevertheless, judiciously phrased neologisms are consistent with Berger's judgment that sociologists have an obligation to create concepts to serve the research purpose at hand. Berger is unconvinced by claims that the established conceptual apparatus known to most sociologists is sufficiently flexible and exhaustive in coverage to be of universal value. His indebtedness to phenomenology as rendered by Alfred Schutz has contributed to making Berger suspicious of the universality of any fixed set of concepts (see, for example, Schutz, 1967).

In *The Homeless Mind* (Berger, Berger & Kellner, 1973), for example, Berger invoked concepts of his own making, or concepts that he had borrowed from relatively obscure sources, in an effort to more effectively capture the complexities and difficult substantive concerns attendant to dealing sociologically with modernization and modernity. His use of these adventitious concepts, multisyllabic locutions like "mechanisticity" and "componentiality," was prompted by Berger's judgment that understanding and comparing the institutions and styles of consciousness in modernizing and modernized worlds is sufficiently difficult to require application of new and different ideas. Unfortunately, however, the result was a collection of neither stylistically attractive nor illuminating terms that occurred much less frequently than one might expect throughout the rest of the book.

On the oddly infrequent occasions in which I encountered one or more of these mystifying expressions I paused, translated it into everyday language or established social scientific concepts, and wondered why Berger had made this sort of translation necessary. Berger writes well enough, with a talent for making unfamiliar and multifaceted ideas manageable, and the neologisms and the concepts they purport to represent are more trouble than they're worth.

This sort of misguided linguistic inventiveness represents the kind of terminological pretentiousness and confusion that makes outsiders suspicious of sociology as an intellectually constructive discipline. This is exactly the sort of ill-conceived conceptual apparatus that Berger himself has dismissed as "the technical dialect for which sociologists have earned a dubious notoriety" (Berger, 1963: vii).

One or another of these newly forged concepts from *The Homeless Mind* (Berger, Berger & Kellner, 1973) occasionally appears in Berger's subsequent work (see, for example, Berger & Kellner, 1981; Berger & Huntington, 2002; Berger, 2011). When they do, they are usually not defined, and *The Homeless Mind* (Berger, Berger & Kellner, 1973) is not cited as a source. Their meanings, for the most part, are more or less clear from the contexts in which they occur, but their use seems gratuitous, and their presence is aesthetically unsatisfying.

No doubt I'm giving too much attention to a handful of essentially innocuous neologisms, but they are so ill-conceived and graceless as to diminish the credibility of Berger's judgment that sociologists should be able to devise effective concepts peculiar to the demands of the research task at hand, and they carry with them the sound and substance of misguided scientism. It's as if Berger were working against himself by doing a really bad job of generating new, ostensibly useful ideas.

Perhaps this is an unduly harsh judgment, and certainly, in principle, development of new concepts tailored to specific tasks may sometimes be a good idea. They do, however, seem to needlessly diminish the value of sociological concepts already in wide use. It also makes one wonder if Berger fails to appreciate his estimable facility in dealing with complex and difficult social phenomena using everyday language devoid of extraneous, quasi-technical terminology. After all, Berger himself has held that most sociology can be effectively presented in ordinary English (Berger, 1963; Berger, 2014).

In addition, in other contexts Berger has shown that he is perfectly capable of parsimoniously producing suggestively useful and linguistically attractive new concepts well suited to a particular project. For example, when working in South Africa as part of a group charged with anticipating what a post-apartheid nation might look like, he was asked to provide a conceptual framework that would guide the efforts of the disparate collection of bright and accomplished policy analysts with whom he was working (Berger, 2011).

Wisely, Berger used *The Social Construction of Reality* (Berger & Luckmann, 1966) as his point of departure. In addition, however, he invoked the term "cognitive maps" to refer to the interests of the various politically active groups with a stake in developments in their nation (Berger, 2011).

"Cognitive maps" strikes me as just the right, self-explanatory concept to capture the conflicting, overlapping, and independent interests involved. There's nothing awkward or mystifying about it and it serves Berger's purpose quite well. It seems that Berger's talent for inventing new concepts and suitably readable neologisms varies enormously from one project to another, with the fewer the better being a very good guideline.

It's interesting to note, moreover, that Berger used the term "cognitive map" in passing, in a non-technical, conversational way in *Sociology Reinterpreted* (Berger & Kellner, 1981: 20 & 23). Perhaps this was a handy turn of phrase lurking in Berger's half-forgotten, passive vocabulary just waiting for the right time to be expressed more formally (cf. Tolman, 1948). The right time came in South Africa while the feudalistic racial policies of apartheid were still in force (Berger, 1986 & 2011).

Moreover, the accessibility of his large body of work long ago persuaded me that conceptual mystification is something Berger is usually determined to avoid. Nevertheless, when Berger later discusses the meaning and genesis of some of these same novel concepts in *Many Globalizations* (Berger & Huntington, 2002) and *Adventures of an Accidental Sociologist* (Berger, 2011), the ones I had found troublesome are no more clear or compelling than in *The Homeless Mind* (Berger, Berger & Kellner, 1973).

Commonsense knowledge and the reality of everyday life

As a contemporary sociologist of knowledge, Berger is primarily interested in what ordinary people know – taken-for-granted knowledge or long-established sets of instructions – rather than arcane philosophical accounts of the ideas of highly specialized scholars. Finding truth according to some presumably infallible, absolute, and fundamental standard is not part of Berger's project. Even his work on religious faith acknowledges its organically developing character, avoiding insistence on rigidly timeless truths totally independent of their formulation in dramatically changing circumstances (Berger, 1967 & 1992).

Instead, his first concern is the reality of everyday life, or commonsense knowledge. Since we live in an intersubjective world, commonsense knowledge is the knowledge that we share with others in our everyday activities and mundane interactions.

Berger insightfully grasps the fact that this is the kind of knowledge without which society could not exist. He recognizes that the reality of everyday life is intersubjectively determined. The shared, all-purpose nature of commonsense knowledge provides an important part of an answer to a question of first importance to Durkheim, Parsons, and numerous other sociologists, namely "how is society possible" (Durkheim, 1897/1997; Simmel, 1910). Given its pervasiveness, versatility, and the ease with which it is shared, the answer for Berger is that commonsense linguistic interaction makes it so (Berger, 1963).

The one best statement of the utility and importance of commonsense knowledge that I have read is George Herbert Mead's account of the development and nature of the generalized other in *Mind, Self, and Society* (Mead, 1934/1967). That Berger reveres Mead's contribution to sociological theory is manifest throughout his work, especially *The Social Construction of Reality* (Berger & Luckmann, 1966).

Whether or not Mead's work could be used to develop a sociological psychology, one that derives its principles and perspective from a sociological understanding of everyday life, as Berger suggests, is a question that, for now and perhaps forever, remains unanswered (Mead, 1934/1967). Raising the possibility of such a development, however, is a tribute to Mead of the sort that Berger reserves for few others.

I think that an important implication of Berger's perspective is that it is a waste of time for the sociologist, rather than the philosophical anthropologist, to struggle with deriving a definition of human nature (Berger & Luckmann, 1966; Berger, 1967). If sociologists are interested in human nature as a foundation for their theorizing and substantive work, I think it's best to construe it as contingent and context specific. We are products of the circumstances in which we live, and our nature is contextually mutable and historically distinctive. In effect, we produce ourselves and are produced as part of the dialectical social process of living with others in varied sets of circumstances that assure that our natures will be variable as well (Marx, 1844/1988; Berger, 1967). As a result, there are as many different human natures as there are distinctive cultural forms.

Berger's emphasis on the sociology of everyday life as people actually live it is an important reason why his work, with occasional exceptions, has been consistently interesting. It also attunes us to the widespread availability of meanings without which no society could exist. It focuses on the relationship between the character of the social context in which we live and what we take to be factual and real.

Berger also makes the interesting observation that if we encounter something that seems not to fit with what we commonsensically take to be correct, we, as a matter of course and largely unself-consciously, try to reinterpret it in a way that makes it understandable, something that is consistent with our experience. This is the kind of insight that fascinated the sociologist Harold Garfinkel when he wrote his influential book *Studies in Ethnomethodology*, a book Berger admires for reasons he does not explain and I don't understand (Garfinkel, 1967).

In spite of my reservations, Berger expresses admiration for Garfinkel's early work. His approval is constrained, however, by recognition that Garfinkel's sociology limits us to the micro-social level. By implication, I take this to mean that Berger's research as a sociological tourist gives him access to the macro-social level, something I had not expected in an approach I had initially taken to be essentially ethnographic and decidedly informal. A moment's reflection on his writings, however, will make clear that this assumption was erroneous. (See for example, Berger, 1967/1990, 1986, 1992 & 2011.) Berger's ethnography, if that's what it is, is an unusual sort that somehow gives him access to macro-social phenomena.

Berger's sociological vocabulary and mode of expression are sufficiently distinctive such that the more of his work we have read the better we understand him. At his best, Berger speaks with a voice peculiarly his own, original, ingenious, and interpretable, but best understood in the context provided by his other work.

Berger's use of the word "nomos" in *The Social Construction of Reality* (Berger & Luckmann, 1966), and elsewhere, for example, may be more or less readily understandable for his inveterate readers. They can compare it to "anomic" or "anomie," ideas borrowed from Durkheim (1897/1997) that Berger uses frequently (see, for example, Berger, 1963, 1967/1990, 1986 & 2011). Those who know that anomie refers to social circumstances where meaninglessness and cultural deregulation prevail might very likely, given a suitable context, correctly infer that a nomos is a social setting where established meanings and cultural stability beneficially hold dominion. For others, however, nomos is likely to be a source of confusion and frustration, especially since it does not appear in most dictionaries.

Even Berger's steadfast readers, however, may have trouble with his use of "nomoi" as the plural of nomos. In addition, when Berger uses "denomoized" to refer to those who have been divorced from the commonsense cultural reality of the nomos and now live anomic lives, and "nomoizing" to mean the building of "nomoi," he may leave still more readers behind. These linguistic inventions used, to my knowledge, for the first time in *The Sacred Canopy* (Berger, 1967) make clear that Berger's penchant for hard-to-interpret, strange-sounding neologisms predated *The Homeless Mind* (Berger, Berger & Kellner, 1973), which we will refer to shortly.

Furthermore, it's worth noting by way of clarification that Berger uses the spelling "anomie" throughout most of his work, though as we'll discuss below he employs the term to mean not only meaninglessness and cultural deregulation, but to cover egoism or rootlessness and lack of belonging as well (Berger, 1963). Moreover, in *The Sacred Canopy* he uses the spelling "anomy" (Berger, 1967). His reasons for inconsistent spellings of the same word, and tacitly attributing alternative meanings to the same concepts are not explained.

Given the contexts in which anomie and anomy are used, however, I take anomie to follow tradition and apply only to collectives – groups or social systems – where an established, inter-related body of meanings has come under assault and broken down. Anomy is sometimes used in the same way, but Berger also employs it in reference to individuals, people whose world has become culturally deregulated – living meaningless lives that are devoid of purpose (Berger, 1967).

Even brilliant scholars inevitably err

While we're addressing issues pertaining to the utility of concepts and clarity of composition, it's useful to call into question what at first glance may appear to be inconsistencies and ambiguities in Berger's account of his work. Were these not all from the same volume, *Adventures of an Accidental Sociologist* (Berger, 2011), I doubt that I would have considered them worth mentioning. However, their simultaneous appearance in the same book might be taken as evidence of carelessness and an exaggerated ease of interpretation that bespeaks an accomplished scholar's self-satisfied lack of close attention to assuring clarity in what he has recently produced. This unfortunate set of characteristics is something that Berger's later work exhibits fairly frequently (see, for example, Berger, 1986, 2011 & 2014).

Using the phenomenological concept typification (Schutz, 1967), a set of common social assumptions and shared meanings that emerge through social interaction, Berger unequivocally asserts that sociologists' concepts and their application must be sensitive to meanings given to social phenomena by those being studied. How a sociologist acquires this knowledge in the first place is an issue that Berger addresses at length in *Reinterpreting Sociology* (Berger & Kellner, 1981).

His treatment, however, has the effect of dismantling the established conceptual apparatus of sociology in favor of ad hoc constructions consistent with the understanding of the subjects of a particular research project. If consistency is not established, the concept is discarded. To do otherwise, according to Berger, is to fall victim to misguided positivism (Berger & Kellner, 1981). In effect, research is directed by meanings shared by people who are subjects of the research. In practice Berger's procedure for concept formation reads like a process that never ends, and can be written up in only a provisional, for-the-time-being fashion, always subject to further development and clarification. This also reflects Berger's growing interest in sociology as an interdisciplinary endeavor. Clearly, this is a labor-intensive process quite different from the way most sociology is done.

More to the point, when invoking Robert Merton's concepts of manifest and latent function (Merton, 1949/1968), Berger repeatedly and persuasively makes the case that innumerable manifestations of the social world are not what they seem to be or what participants say they are. Thus, a primary concern of the sociologist, one that sets him apart from other social scientists, is to look beyond and beneath manifest claims to find latent truths and meanings (Berger, 1963).

These two judgments, the need for concepts to be consistent with participants' typifications, and the priority of a sort of theoretically neutral debunking, need not be contradictory. But the two are juxtaposed so conspicuously that it's difficult not to suspect an inconsistency (Berger, 2011). For the sake of clarity, Berger might acknowledge and reconcile them rather than leaving it for his readers to figure it out for themselves. It's not difficult, but the need to do so may undercut a sense of certainty and completeness in Berger's later work.

In another troublesome instance, after Berger explains what he trumpets as his "discovery of capitalism," he makes a truly peculiar admission. In his words, "it was no longer possible for me to deal evenhandedly with capitalist and socialist modes of development" (Berger, 2011: 142).

The superiority of capitalism refers not only to enhanced productivity, but to what Berger terms "the calculus of pain" – minimizing the toll of human suffering involved in modernization – and the "calculus of meaning" – deference to the value-laden worldview of those afflicted with modernizing processes (Berger, 1986: 12). Capitalism, in most instances, proves superior to socialism, as Berger sees things.

But this is the same Berger who earlier had taken pains to explain that the substance, theory, and method of sociology enabled sociologists to divest themselves of preconceptions, preferences, and prejudices while providing genuinely objective information. To do so they used the relevance structure, or means of deciding what information is pertinent and what is not, peculiar to sociology (Berger & Kellner, 1981). (The term "relevance structure" is very roughly comparable to Kuhn's concept "paradigm" as presented in *The Structure of Scientific Revolutions* [Kuhn, 1962], and Althusser and Balibar's "problematic" as unreadably discussed in *Reading Capital* [Althusser & Balibar, 1970]). This is also the same Berger who in *Invitation to Sociology* took the positon that for "a sociologist there is one fundamental value only – that of scientific integrity" (Berger, 1963: 5).

It's difficult to imagine that the superiority of capitalism for all times and all places had become so overwhelmingly certain that Berger, the value-neutral sociologist, no longer needed the instruments and safeguards of his own social scientific discipline to avoid misinterpretations, misconceptions and bad decisions. However, Berger did not acknowledge what appears unmistakably to be an inconsistency in his presentation. In this instance, moreover, the contradiction is so obvious that it cannot be dismissed as an incidental oversight or misguided nitpicking, something that could easily be explained away (also see Berger, 1986).

One wonders if, for reasons not recognized or simply not given, Berger had abandoned his role as a professional sociologist guided by the substance and method of his discipline. In view of its consequences in terms of lethal instability in the Middle East, I think one might be excused for being suspicious of Berger's objectivity when he characterizes arch-capitalist George W. Bush's foreign policy as merrily "adventurous" (Berger, 2011: 204) – like a kid with a new toy that soon will be broken. The nature of Berger's involvement with the father and son Bush Administrations took me completely by surprise, and cast additional doubt on the objectivity and evenhandedness of Berger's work (Berger, 2011).

Furthermore, when Berger argues that capitalist efforts at modernization are superior in manifold ways to socialist endeavors, his presentation is unclear as to whether capitalism inheres in the nature of a development project, in the organization or nation that in one way or another sponsors it, or in both. The same applies to socialist undertakings. As a result, Berger tacitly denies the variability that exists within broad categories and holds that in every instance one dichotomous variable, capitalist/socialist, somehow covers all differences with regard to the effectiveness of each and every modernization initiative (Berger, Berger & Kellner, 1973). That this has not been the case is attested to by research and evaluation reports on a wide variety of developmental actions (Hickling-Hudson & Klees, 2012).

I admit to being troubled by Berger's unqualified assertion that a sociologist who is adequately sensitive to what and whom he is studying must devise his own concepts, not relying on those already available (Berger, 2011). If so, his travels as what he terms a sociological tourist must be intellectually exhausting. The sort of creativity required by the fabrication of new and useful concepts for every research endeavor, including those that involve just sitting, watching, and thinking is made manifest only through extremely hard intellectual labor.

Oddly enough, however, Berger then makes the case that the sociologist differs from other observers in that his frame of reference is rich in the research and theory that constitutes the body of knowledge, some of it due to Berger himself, and the peculiar approach to understanding social phenomena that give sociology credibility as a roughly codified and worthwhile discipline (Berger, 2011). Once again, we have what I take to be awkwardly ill-fitting assertions that go unacknowledged and unreconciled.

Insofar as sociologists' established conceptual wherewithal is of real and comprehensive value, as Berger asserts, one seems justified in assuming that his conceptual requirements have already been satisfied. Are sets of meanings and institutional structures really so different from time to time and place to place that a generally applicable body of concepts is not already available and pretty well adequate to most sociological tasks?

If Berger's artless effort at forging original, case-specific tools in *The Homeless Mind* (Berger, Berger & Kellner, 1973) is taken seriously as an example of this process, we are indeed faced with an ugly and troublesome set of circumstances. If Berger's application of Weber's 100-year-old Protestant Ethic to the broad range of disparate countries that constitute East Asia and then, an ocean away, Jamaica, is accepted as suitable to today's world, it seems that the usual repertoire of time-worn concepts may be adequate after all (Roth & Wittich, 1978; Berger, 1992, 2011 & 2014).

Perhaps I'm finding inconsistencies where none blatantly exist. Nonetheless, I'll offer a final example that I find troubling. Mediating structures, also known as intermediate or secondary groups, are effective – perhaps essential – in coming to terms with a variety of social and cultural needs in large, complex, modern societies (Putnam, 2000; Berger, 2011). Enabling their members to overcome the sometimes-debilitating family income limitations intrinsic to minimally regulated capitalism, however, is something else again, something that fits at best awkwardly in the mediating structures program and receives almost no discussion, certainly nothing of importance (cf. Berger, Neuhaus & Novak, 1994).

How mediating structures as fostered and implemented in Berger's world would accomplish an economic objective is unclear. As discussed in Berger, Neuhaus & Novak (1994), mediating structures serve to maintain norms, values, and social and cultural stability when bounded areas are threatened by the ideational depredations of megastructures, whether public or private.

As such, they are the "value generating and value maintaining agencies for society" (quoted in McClay, 2009). The limitations inherent in this characterization, however, do not prevent Berger from proceeding as if mediating structures would not only empower citizens but, in an unlikely turn of events, enable them to secure the individual and familial economic wherewithal essential to living a decent life. Oddly, however, after making a hopelessly vague reference to placing a "floor of decency under everyone" (Berger, Neuhaus & Novak, 1994), Berger makes what appears to be a self-serving declaration that finds government-sponsored income maintenance strategies beyond the scope of his discussion and thus, one might infer, beyond the boundaries of mediating structures. Berger's account then devolves into a brief and conventionally conservative treatment of school vouchers, followed by a painfully uninformative reference to care of the handicapped and disabled.

Nevertheless, in his discussion of the role of capitalism in modernization among countries in East Asia, it is plainly obvious that Berger gives priority to providing prosperity for individuals and families. Living the good material life supersedes all else! Whether this occurred in a democratic society or an autocratic or dictatorial one was of little or no concern (Berger, 1986 & 2011).

In my view, when Berger proceeds to effectively minimize, obscure, and garble the nature and importance of individual and family income in the United States, while emphasizing its gravity in East Asia, he deposits another inconsistency for readers to sort out for themselves. It seems entirely too easy, and perhaps simply unfair and unduly oblique to attribute this to Berger's sociologically unseemly commitment to capitalism. That, however, is an explanation that Berger himself has unmistakably invited (Berger, 1986; Berger, 2011).

Surely, *Adventures of an Accidental Sociologist* (Berger, 2011) does not embody every ill-considered judgment Berger ever made, and not all are of recent vintage. All research methods, laudable efforts at sociological tourism emphatically included, have their limitations. Sometimes, moreover, the limitations are difficult to see, and sometimes sociological tourism itself may manifest aberrations we had not expected, as when a sociological tourist loses his way.

In Berger's chapter on "The Problem of Theodicy" in *The Sacred Canopy* (Berger, 1967), he apparently again is engaged in sociological tourism, though without his usual methodological certitude. It seems clear that he gets lost in ungrounded theorizing and proceeds blindly without the benefit of observation, whether rigorous and sustained or lax and dependent on happenstance, or any other kind of empirical source (Berger, 1967: 55–80). When that happens, Berger is reduced to relying on speculation, something that he doesn't do very instructively.

When Berger argues that man cannot accept aloneness, and that he cannot accept meaninglessness, he is tacitly invoking Durkheim's concepts of egoism, or social deracination, and anomie, or cultural deregulation. That's all to the good. But embedded in an argument more convoluted and lacking in substantiation than perhaps anything else the usually more plain-spoken Berger has written, he offers two trite and paradoxical aphorisms as if they were well-founded theoretical contributions, somehow pertinent to fully understanding egoism and anomie: "It can be sweet to suffer pain at the hand of one's lover," and "It can be sweet to be punished for the sovereign authority of the state" (Berger, 1967: 55).

Fatuous nonsense such as this belongs, side by side, with simple-minded ironies such as "You always hurt the one you love." Berger has taken the position that theodicies or efforts to explain why evil persists in spite of the reign of an all-powerful and thoroughly good God are inevitably irrational, hence the practical absurdity, though not the soft-core comic book silliness, of the two aphorisms. It is clear that when circumstances force Berger to substitute naked speculation for sociological tourism his presentations may be weak, even vacuous.

For one who sought to make his case using the concepts masochism and sadism, however, these paradoxical propositions may seem to make a good deal of sense. So perhaps it's best not to found our theodicies on two such practically improbable ideas.

Whether or not that is good advice, we can see that Berger is not at his best when he proceeds as a speculative philosopher rather than a concretely informed sociological tourist. Nevertheless, Berger is capable of high-quality theoretical work, as we see in *The Social Construction of Reality* (Berger & Luckmann, 1966) and in much of *The Sacred Canopy* (Berger, 1967). His lifelong commitment to sociological tourism, however, seems to limit him, for the most part, to doing his best sociology, including his best sociological theory, when informed by even a relaxed empirical component.

Berger's discovery of capitalism

I haven't been able to pin down exactly when Berger discovered capitalism and was so deeply impressed that he was rendered unable to evenhandedly compare capitalist and socialist efforts at modernization or anything else. However, Berger acknowledges that at least as far back as 1969, four years before publication of *The Homeless Mind* (Berger, Berger & Kellner, 1973), he was sensitive to the possibility that socialism might provide a better program for modernization – a program of institutional change and transformations of consciousness – than capitalism (Berger, 1986; Berger, 2014).

Nevertheless, during the nearly twenty years between *The Homeless Mind* (Berger, Berger & Kellner, 1973) and *The Capitalist Spirit* (Berger [Ed.], 1990) the sum total of Berger's scholarly experience came together and persuaded him that capitalism, whatever its source and however variable its manifestations, provided the superior set of practices to, among other things, promote modernization with a minimum of suffering along with acknowledgement of the integrity of Third World values.

At various places throughout his work, Berger presents his discovery of the superiority of capitalism with the same certainty and intensity as the realization of the centrality of unclehood in determining one's identity in a matrilineal society (Berger, 1967). This observation may seem unduly satirical, but it's not meant to have a satirical character. Instead, it reflects a misguided commitment so strong that it divorces Berger from the most important professional constraints that make a competent sociologist (Berger, 2011). This is a good deal more troubling than satirical.

Berger's brief introduction to his edited volume *The Capitalist Spirit* (Berger [Ed.], 1990) is an unmitigated paean of praise to capitalism, the less regulated the more productively impressive (Berger, 1986). Lesser authors might be embarrassed by his lack of restraint and the near complete suspension of his critical faculty. Berger's primary objective in the introduction's five pages is to call attention to the fact that religious organizations' objections to capitalism focus almost exclusively on economic distributional issues, with the poor otherwise left out or short-changed.

In *The Capitalist Spirit* (Berger [Ed.], 1990) Berger wants to shift emphasis from distribution to the massive productivity gains that come with the development of capitalism. Though he doesn't put it in just these words, I take him to mean that issues of equitable distribution and widespread comfortable subsistence can be realistically and consistently raised only because capitalism is capable of producing an enormous surplus. On that basis, he enthusiastically opts for capitalism.

Given that Berger is a scholar of extraordinary erudition, I'm sure he knows that Marx himself took the position that a fully developed industrial economy was a necessary prerequisite for socialism to succeed (Marx, 1867/2011). Even if Berger is convinced that socialism would fail under any and all circumstances, his credibility, in my view, calls for him to acknowledge that his references to the wreckage ostensibly wrought by socialism invariably refer to social systems that started off lacking in Marx's essential prerequisite for thoroughgoing capitalist development.

I think that even readers favorably disposed to Berger's position, if they thought about it, might agree, if only to enable Berger to avoid seeming mean-spirited. Judgments such as "the … insouciant identification of Christian morality with capitalism" (Berger [Ed.], 1990: 3) as used by Berger are either indecipherable or ugly, perhaps both. His heavy-handed dismissal of organized religion's endorsement of intrinsically economic efforts to promote equity and eliminate poverty could have been written by Friedrich Hayek (1944/2007). Berger may take that as high praise, but if I were him I'd accept it as such only if I were eager to be characterized as a stubborn reactionary with brutal friends, people such as Augusto Pinochet, in high places.

Value-free sociology

In any case, I think we can agree that however one interprets the preceding examples, including the discovery of capitalism, that I find troublesome, the sociologist as sociologist can provide only the conceptual wherewithal needed to demonstrate that alternative interpretations are possible and, in an epistemological sense, perfectly legitimate. Value judgments, however, remain exclusively within the purview of those not active as sociologists (Berger & Kellner, 1981). This is an assessment that Berger makes repeatedly throughout his work (Berger, 1963, 1986, 2011 & 2014).

Understanding the practical implications of a prevailing normative framework is well within the accomplished sociologist's capabilities, as with the precarious persistence of traditional ways of behaving and understanding in a society undergoing a process of rapid modernization. In fact, according to Berger, understanding the radical change intrinsic to what we today call modernization gave rise to sociology as a discipline and social theory as an effort to provide a foundation for that emergent social science (Berger, 2014). However, discerning likely outcomes and making empirically and theoretically informed judgments is as far as the sociologist may legitimately go (Berger, Berger & Kellner, 1973). Questions of value, as we have seen, are outside his scope.

Minor mystifications

It's a minor matter, of course, but I was surprised when Berger invoked the notions "politically correct" and "politically incorrect" (Berger, 2011). I'm not suggesting that Berger should be linguistically formal, dense, or pretentiously convoluted when that is neither necessary nor judiciously fitting, but use of politically correct terminology suggests adherence to a right-of-center ideology and does so in an offhanded way.

As far as I can tell, when Berger takes a position on an issue, it is rarely for reasons that might be characterized, as in this instance, as offhandedly thoughtless, even by those who disagree with him. (The sadism-masochism examples reported above are noteworthy exceptions.) In his more recent work, especially, he may rightly be judged as occasionally careless, wrong-headed and prejudicial, but rarely offhandedly thoughtless. I think that what really bothers me is my conclusion that Berger trivializes his work when he resorts to inherently meaningless, over-used, and rather stupid locutions and talk-radio turns of phrase.

I wish, moreover, that Berger were less inclined to make an important assertion, and then follow up with a sentence that begins with "Put differently" This hardly seems worth mentioning, but when I was a graduate student, apparently trying to emulate material I had read by Berger, a scholar I held in high esteem, I found myself adopting the same usage, and no on liked it. I must admit that its repetition is annoying.

Much the same applies to *mutatis mutandis*. I'd be willing to bet a tidy sum (if I had one) that Berger uses this phrase more frequently than any other author in print.

A provisional summing up

Nevertheless, if we look past these occasional missteps into the labor-intensive creation of concepts with ill-advised designations, hackneyed conservative catch phrases, excessive use of Latin admonitions, and what I take to be unexplained inconsistencies, Berger's presentation is quite forceful. He's definitely an accomplished sociologist of substantial intellectual attainment and he writes as such. His prose flows smoothly, carrying the reader along, minimizing the inevitable effort that comes with reading inherently dense material that, especially for the beginner, is conceptually difficult and sometimes counter-intuitive. Berger also makes judiciously effective use of wit and humor.

Whatever differences I may have with Berger, his brilliance and productive sociological imagination are well known to those of us first introduced to the socially constructed nature of our world by his exemplary work (Berger, 1963 & 1967/1990; Berger & Luckmann, 1966; Berger & Huntington, 2002). It takes time and effort to appreciate and fully come to grips with his remarkably astute and fertile observations, but they produce a radical change in the way that those of us starting off with nothing but the hackneyed perspectives and dull prospects of everyday life understand the world and our place in it.

Inevitably, it yields a transformation, both salutary and troubling, in what Schutz termed our "life world," meaning the world as we experience it (Schutz & Luckmann, 1973). For sociologists, the socially constructed point of view has become an essential and typically taken-for-granted part of their relevance structure or selectively defined perspective, determining what they see and how they see it, though this is something many have forgotten and fail to appreciate.

I imagine that some would find it awkward and perhaps a bit tawdry to be associated in any resolute way with a colleague who relies on a procedure he terms sociological tourism to describe the methodological nature of most of his research. If for some of them sociological tourism necessarily bespeaks an embarrassing lack of rigor, I think they may be indulging in misconceptions as to just what rigor may mean.

It may not be immediately evident, but a socially structured world, one that could be assembled in an indefinitely large and varied set of ways, opens the door to the difficulties posed by the assumption of a thoroughgoing relativism. In this way, Berger has shown us that there are no verifiable eternal absolutes, at least not in the temporal realm, and social arrangements of all kinds, including their foundations in a specific moral code, are, in truth, transitory (Berger, 1992; Berger & Huntington, 2002).

As Berger's work with modernization and modernity makes concretely clear, each of us lives in a world best characterized as consisting of precarious institutions and correspondingly fragile structures of consciousness. This world is quite real, heavy-laden with objective facticity, but it may prevail only for the short term, and it provides numerous vantage points, each revealing a different position as to what is truth (Berger & Luckmann, 1966). If all truth is not socially relative, it is definitely socially relational: The more vantage points to which one has access, the more persuasive one's claim to judging what is true and what is not true.

Yes, there are innumerable vantage points, including those available only to highly specialized theorists whose work is foreign – perhaps completely unknown, intellectually inaccessible and of no concern – to the ordinary people who make up the massive bulk of any society. It is very much to Berger's credit that he has not followed disciplinary precedent and restricted his work to the output of specialists producing thoroughly particularized knowledge of no practical value or interest to most people. The institutions that produce this kind of knowledge have their place, but not in a treatment of the sociological understanding of knowledge that would even very roughly approximate comprehensiveness and be of general interest (Berger & Luckmann, 1966).

Furthermore, as we will discuss below, Berger holds that thoroughgoing relativism and hard-and-fast fundamentalism are both unfortunate manifestations of the same process of modernization. This paradoxical claim follows from Berger's judgment that both uncompromising relativism and unyielding fundamentalism are inherently inconsistent with civil or scientific discourse, an inconsistency that Berger judges to be peculiar to the modern world (Berger [Ed.], 2009).

One can readily see the difficulties posed by relativism when one posits that there is the executioner's narrative and the narrative of the executed. While both statements may be accurate and informative, hideously graphic substantive disparities between them force the relativist to face the empty horror covered by his relativism. The shock may be so overwhelming that the relativist is jolted into a position of fundamentalism, where some themes, ideas, and acts become either unambiguously demanded or forever denied (Berger [Ed.], 2009).

Berger's position is that the vast majority of people, without extended internal arguments or much thought at all, unself-consciously gravitate toward a loosely defined middle position (Berger, 1992). For better or worse, and without substantiation, Berger argues that this was and remains the natural social path of people in traditional societies. Berger may be right when he takes the position that departures from convention are regarded as eccentric and harmless oddities among traditional people, but this seems inconsistent with his embrace of pluralism as the precursor of secularism, and sharply at odds with the work of Durkheim and Tönnies (1887/2001).

Nevertheless, for the sake of clarity and completeness, it is useful to acknowledge that most conspicuous and influential early proponents of modernization theory were also closely tied to what is commonly termed secularization theory. Secularization theory is based on a simple, seemingly commonsensical idea: As a society becomes more modern, it becomes less religious. To scholars as varied as Marx, Durkheim, Freud, and Weber, this seemed self-evident.

They saw the modernizing world becoming increasingly rational and science-based, and the decline of the sacred in the midst of the secular seemed an inescapable development.

According to Berger, however, the intellectually untenable character of unmitigated secularization theory became obvious in the 1970s (Berger [Ed.], 2009). Instead of becoming less religious, much of the world at large was becoming substantially more religious, at least with regard to membership and attendance at religious services. This was not just a vague impression, but a social fact that could be verified using simple summary statistics attached to ambitious data collection efforts. In Berger's view, "modernity … is comfortably compatible with religion of one kind or another" (Berger [Ed.], 2009: 3). As is too often the case, this is an interesting judgment that Berger offers unequivocally but does not develop.

Nevertheless, Berger acknowledges that modernization as manifest in secularization theory does tend to undermine established beliefs and values that have withstood the test of time. Then, however, he admonishes us to avoid confusing the effects of modernization with those of pluralization. Modernization, in Berger's view, may or may not secularize, but it almost certainly pluralizes. The primary and uncomfortable consequence of pluralization is that in a modernized world, "certainty is much harder to achieve" (Berger [Ed.], 2009: 6).

The modern world is a varied one that presents us with the opportunity and need to make choices. This betokens freedom from confining traditional norms and institutions, but it also means a proliferation of plausibility structures, and diminished guidance and support in selecting one way of worshipping, or not worshipping, believing or not believing, over another.

Berger uses the conceptually powerful but easily misinterpreted term plausibility structure to refer to a group of like-minded, affectively kindred people who commonly bear witness to a distinctive faith, creed, world view, or symbolic universe (Berger [Ed.]. 2009). The durable stability of a specific religious faith, for example, is dependent on the existence of a cohesive group with members who share its tenets and interact together accordingly. Furthermore, it seems clear that any commitment that might take the form of a belief, religious or not, presupposes the existence of a plausibility structure. Long-time soldiers with a strong and durable commitment to each other and their battle-tested brigade constitute a plausibility structure (Blum, 2002).

Berger's commitment to pluralism as an explanatory factor with regard to ascendance of the secular and the diminution of the sacred is based on the claim that a more plural population will be sufficiently diverse to undercut the formation of existing plausibility structures and foster the emergence of new and varied ones. Undercutting plausibility structures means weakening the foundations for faithful commitment to a specific set of ideas, values, or convictions (Berger, 1992). As Berger sees it, diminishing the strength of one plausibility structure marks the beginning of the proliferation of plausibility structures and the religious faiths and other claims made with conviction that they support (Berger [Ed.], 2009). Whether or not this pattern of relationships is convincing is something we will discuss further.

For the time being it's instructive to acknowledge that, sharply at odds with secularization theory, the world we share has occasioned a dramatic growth in religious practice and formal organization. Western and Central Europe are the most conspicuous exceptions to this development, but elsewhere modernization and modernity have not had the expected secularizing consequences.

It seems useful, therefore, to imagine how the foregoing account might lead to an increase in the incidence of fundamentalism. As noted above, in a modern world, characterized by diversity with regard to just about everything of importance, certainty is hard to achieve. In the absence of certainty with regard to religious faith, anomie is a painfully likely outcome (Durkheim, 1897/1997). Fundamentalism, whatever its limitations, is a hedge against anomie in a modern world suffused with pluralism.

Relativism, following Berger, is something that varies with the domain or area of human activity with which we are concerned. Since science gives us conditional propositions that can never be validated once and for all, relativism of a sort is inescapable. One proposition, in principle, is always open to being rejected in favor of another.

Moral relativism, however, is another matter, though I don't think Berger's case is a strong one. Nevertheless, he holds that some things – torture, slavery, and serial rape, for example – are so thoroughly repugnant that almost all of us are sure to reject them without reservation. Otherwise, relativism is just as objectionable as fundamentalism (Berger [Ed.], 2009).

The trouble with a revulsion hypothesis such as this as a foundation for relativism, however, is that it's difficult to think of one that is not peculiar to a specific time and place. The examples suggested above would seem surely to apply to the modernized West, and yet Abu Ghraib was run by and for the U.S. Army and the Central Intelligence Agency (Danner, 2004). Furthermore, much of the rest of the world accepts what many of us take to be moral outrages as perfectly suitable, even culturally essential. In this instance I think that Berger has lost sight of his value-neutral take on sociology.

At this point, I think it's useful to render the judgment that Berger has not presented evidence sufficiently persuasive to prefer pluralization theory to secularization theory. He has devoted more attention to pluralization theory, but it seems likely that this is because Berger was convinced ahead of time that it has greater explanatory power, and he has concluded that secularization theory has been empirically falsified (Berger, 2014), even though the evidence he offers to support this contention is substantively thin. Perhaps this is one reason why Berger is especially interested in a social and theoretical middle ground to which most people unself-consciously gravitate. This is the position of greatest social and individual well-being, avoiding the violence of fundamentalism and the pointlessness of relativism.

Nevertheless, Berger acknowledges that in his early years as a sociologist he was a proponent of secularization theory. With regard to the demise of secularization theory, given a close reading of Berger's essays in *The New Sociology of Knowledge* (Berger, 2014: 43) he seems equivocal. On the one hand, Berger holds that "a whole body of literature by historians and sociologists loosely labeled 'secularization theory, 'is essentially mistaken." On the other hand, he argues that "some of the writings he produced [based on this perspective] still stand" (Berger, 2014: 43).

I suppose it occasionally occurs that work of lasting value can be written based on what may be a faulty premise such as secularization theory. When making the case that secularization is intrinsically impaired and must give way to an alternative, in this case pluralism, compelling examples would be quite useful. Too often throughout the broad span of his multifaceted work, Berger makes interesting observations and fails to develop them. He gives the impression of being one who is satisfied with writing clever aphorisms but lacks the interest to follow through.

Doing this repeatedly can only diminish the credibility of his offerings. The following makes this case clearly: "To be sure, modernization has had some secularizing effects, more in some places than others" (Berger, 2014). Again, the appearance of waffling could easily be dispelled with a well-chosen sentence or two.

That said, I was stunned when Berger juxtaposed conversations with angels alongside the findings of particle physics. I know little or nothing about either, but until I read *A Far Glory* (Berger, 1992), I was convinced that the existence of one was necessarily inconsistent with the insights of the other, and that conversations with angels were manifestations of psychosis or topics for fictional accounts of the supernatural or metaphors whose meaning escapes me.

Berger, however, to his credit and our benefit, thinks it's more likely that both may exist in an interpretable form rather than one or the other. In my experience, it is the rare, courageous, and wise social scientist who acknowledges the possibility of angels or other supernatural manifestations, much less conversing with them while also studying contemporary physical science. Given his stature and the quality of his work, Berger, in effect, grants the rest of us the freedom to believe in that which is sacred and to worship in ways long since written off by intellectual agnostics and atheists. Broadly speaking, much of the best that Berger has written is consistent with a sustained endeavor to join spiritual aspirations with the worldly social context in which one lives (Berger, 1967 & 1992). An ambitious project to be sure.

2 Institutions: Berger's conceptual foundation

Furthermore, as the world becomes more modern, it is more thoroughly infused with an ethos of instrumental rationality. This may leave many feeling lost, confused, and deeply uncertain, yearning for anchorages and the stability of the sacred. Such concerns have profound implications for the commonsense knowledge of everyday life. They will become much clearer as we proceed. For now, we will content ourselves with the observation that institutions and the process of institutionalization are fundamental to doing and understanding sociology, giving substance and meaning to its prose, presentation, and purpose. Discussion of institutions in Berger's work is extraordinarily illuminating, sometimes dense and sometimes necessarily counterintuitive, but still acutely informative (Berger, 1967).

The study of institutions is loaded with unexpected insights. For example, one unexpected characteristic that I find especially interesting about any collection of institutions that provides the organizational wherewithal of an integrated social system is that they need not be somehow logically compatible. Logical consistency seems like something we should be able take for granted with regard to a set of institutions in the same social setting, but following Berger, this need not be the case (Berger, 1963).

Berger's position is that the apparent logical compatibility of institutions is largely a product of memory born of common experience. In effect, it has a biographical nature whereby human beings impose a sense of logical consistency on a diverse set of institutions where such consistency is not available to a detached observer (Berger & Luckmann, 1966). After all, social institutions are ongoing social creations, things that human beings make and continue to maintain. It would be odd, indeed, if the institutional order, whatever its affirmative character, were actually perceived and, as a practical matter, understood as being made up of elements that human beings found conflictually at odds with each other.

It's worth noting, however, that Berger's discussion of the autonomy of institutional types is presented in an atheoretical, apolitical manner. One wonders if, for example, most Marxists would accept the idea that the institutions constituting a socialist society could be logically independent, not subject to the same overarching, thematizing framework. For example, lack of fit between the forces of production and the social relations of production would betoken serious trouble as Marx saw things (Marx, 1867/2011). Perhaps a Marxist would invoke the once-ubiquitous and often misused notion of relative autonomy in fabricating a response to this issue (Poulantzas, 1978/2015). Relative autonomy, however, is an idea that has proved to be conceptually problematic (Jessop, 1987), and it has yielded heated disagreements among otherwise compatible theorists.

More to the point, however, there is an enormously broad range of activities that could be institutionalized, rendering participants' behavior predictable and taken-for-granted. Sexual behavior and the division of labor may each be institutionalized, but is there any reason why we should expect logical compatibility? I'm not even sure what it would look like. In short, the felt need for logical compatibility among institutions is premised on the assumption that societies are much simpler than they actually are, and that they come unglued if the institutions are not demonstrably compatible. The example above, dealing with forces and relations of production, suffers from precisely this sort of over-simplification, though otherwise it may have merit.

It could also be that Berger would respond to this issue by asserting that the process of institutionalization he is describing pre-dates the imposition or acquisition of a particular social type. Slave society, feudalism, capitalism, socialism may all be in the distant future, as human beings participating in an institutionally rudimentary society, one that is just taking shape, begin to think about ways that essential tasks could be done better or in a manner that more adequately serves specific social interests. The externalization, objectification, and internalization intrinsic to the creation and maintenance of institutions peculiar to a more or less well-developed society could then begin to take full form (Berger, 1967).

These closely interrelated and simultaneously occurring processes – again, externalization, objectivation, internalization – constitute the basic dialectical configuration whereby, following Berger, the social world is created by man, and man creates himself. We will encounter these processes frequently throughout this manuscript, and it is useful at this juncture to give this fairly difficult material a brief explanation, a precursor to a fully developed and instructively illustrated discussion that we will offer later.

Berger construes externalization as the "ongoing outpouring of human being into the world, both in the physical and the mental activity of men" (Berger, 1967: 4). This statement, I think, may be ambiguous, because Berger also emphasizes that human beings are instinct-poor and come without predetermined behavior patterns. So, what do they have to pour out?

However, we can begin to dispel the ambiguity by acknowledging that for Berger human activity is inevitably social activity. As a result, the "outpouring of human activity" is inevitably the outpouring of social activity, human beings responding to each other in ways that bespeak common basic needs and endeavors to meet them. This is the beginning of the patterning of human behavior that is externalized, made manifest, and given meaning in institutions, and we have begun to rid the first term in Berger's dialectic of its ambiguity.

Berger uses objectivation to betoken the process whereby externalization results in the creation of physical or mental products that take on the character of objective facticity. They are instances of the social process of externalization as it takes concrete and independent form. Those who give life to the substance of externalization work through that process to bestow objective facticity in the form of objectivation.

Objectivation, rendering something capable of being independent of its origin, is not a difficult concept to understand, but it takes forms that I had not anticipated. For example, language, perhaps in the case of an assertion about formal education and its present strengths and limitations, exists in a form that is independent of its worldly referents. As such it is objective and informative from one person to another. It also is ideally suited to become part of our common stock of knowledge, where it remains accessible and is part of a repository of commonsense or intersubjectively shared information.

The third part of the dialectical relationship is internalization, which results in the absorption of the products of externalization and objectivation into constructions of the subjective consciousness. Internalization is the part of the dialectic that shows us that while men make society, society in turn makes men. This gives completeness and closure to the dialectic, and we see that without a mechanism such as this, social life and men as social beings would not be possible.

Berger summarizes the dialectic as follows:

> It is through externalization that society is a human product. It is through objectivation that society becomes a reality *sui generis.* It is through internalization that man is a product of society (Berger, 1967: 4).

It took me time and numerous re-readings to come to terms with Berger's dialectic. Now it seems self-evident.

The necessity of culture

Initially I didn't take the dialectic seriously and construed it to be just so much innocuous ruffles and flourishes in a broader discussion of the genesis and nature of institutions. Nevertheless, in due course the dialectic came to make a good deal of sense, and it helped me better understand Berger's perspective. Among other things it forcefully reminded me that man is not born into a prefabricated world but must socially create himself and a world for himself. The nature of that world, according to Berger, is best captured by the word *culture.* Culture provides the substance, structures, and processes that, in contrast to instinct-rich animals, man is born without (Berger, 1967).

While use of the word culture in this way makes sense, I think it might produce confusion for some readers. Culture has so many referents exhibiting different meanings, that it would be useful if Berger could think of alternative terminology. Nevertheless, if we understand culture as reasonably intended by Berger, man's instinct-poor, world-creating nature makes use of the concept culture convenient while also giving it a suitably specific meaning (Harrison & Huntington, 2001).

As the mainstay of a man-made environment that is necessarily involved in a constant and precarious process of being remade, culture, in comparison to the constraints and facilitations of the natural world, is relatively unstable. Thus, even though culture shares the objective, institutionalized facticity of what we call the real world, it is not part of the natural world. To think otherwise is to place one's self in danger of being trapped into reification (Berger & Luckmann, 1966). I think, however, that avoiding reification is much easier than understanding the processes that yield culture and the life it makes possible.

Furthermore, too often we fail to distinguish between culture and society, perhaps using the words interchangeably. However, if we think about it now that we have construed culture as a world created by man for man, we soon see that society is but a specific instance of non-material culture. Culture, in effect, subsumes or incorporates the structures that constitute society and its institutions. This is something that we will discuss further as we proceed.

The ubiquity of rationality

One of the mixed blessings of the modern world is the ubiquity of functional rationality, or as Berger puts it following Weber, *rationalization* (Berger & Kellner, 1981: 14). Modernization, as Berger explains, brings rationality in many forms. Science-based technology itself is unthinkable without the unquestioned ascendance of functional rationality over myth, fable, poetic imagery, and other ways of making sense of the world that pre-date, and to some extent co-exist with rationality (Santayana, 1967).

Rationality is also inherent in modern bureaucratic organization with its category for everything and everything with a rationally justifiable category (Gerth & Mills, 1958). In addition, capitalism, whatever its virtues and abuses, is also thoroughly infused with an ethos of rationality.

Marx himself understood this as early as *The Economic and Philosophic Manuscripts of 1844*, making the case that capitalists were necessarily caught up in the rationally calculable, intensely competitive pursuit of profit to the exclusion of all else (Marx, 1844/1988). Rationality, if wholly focused on things we take for granted in everyday life, is a powerful debunking force, but it can also leave us lost and confused, devoid of anchorages and without non-ephemeral meanings (Adorno & Horkheimer, 1944/2007). Perhaps this explains why Berger's juxtaposition of conversations with angels and particle physics is a welcome, even if unlikely, pairing of supernatural and natural phenomena.

Given the foregoing and to avoid misunderstanding, it would be a mistake to assume that all rationally wrought and purposefully developmental efforts have been put forth by capitalists (Berger & Kellner, 1981). Following Berger, there have been at least three categories of actors putatively promoting salubrious development: capitalists, socialists, and seemingly "non-denominational" modernizers (Berger, 1986). Furthermore, we have already emphasized that in broad categories such as these, a great deal of institutional variability is inevitable.

There is little in the way of systematic comparison of the endeavors of the three groups, but the modernizers seem, in practice, to be indistinguishable from capitalists, certainly this applies to their commitment to rationality in a way that some might hold ignores equity (Berger, 1986). We might be tempted to call the modernizers "unaffiliated technocrats," but technocracy itself is an ideology that, to one degree or another, infuses all developmental efforts (Berger & Kellner, 1981; Veblen, 1904/2013; Wood, 2014).

The sacred as a complement to the rational

Fittingly, Berger makes an extremely interesting observation when he claims that rationality is a primary source of precariousness in the modern social world. Rationality betokens reflection, and the more we reflect, the less we take for granted. Thus, thoroughly institutionalized meanings and practices are subjected to the sort of scrutiny that would seem pointless, and perhaps even sacrilegious or amoral, in a less-developed world, one that has not gone through a process of far-reaching modernization.

Insofar as rational examination in terms of desired end states for a modernized social system reveals deficiencies, taken-for-granted – even nominally sacred – phenomena may be subjected to wholesale revision. Sociology itself does not enable us to make ultimate judgments as to value, but it provides the tools that enable us to do so more rationally (Berger & Kellner, 1981).

With regard to the nature of that which is unself-consciously taken-for-granted, Berger makes an insightful observation, the sort of thing that seems perfectly obvious in retrospect but manifests brilliance when first presented. Specifically, "taken-for-grantedness is maintained in consciousness ... [as a] result of social consensus in an individual's environment" (Berger & Luckmann, 1966).

Again, with good reason, Berger is tacitly invoking the remarkably useful concept of "plausibility structure." He is also suggesting that the sacred, including religious doctrine and beliefs, are subject to cross-fertilization, competition, and even invalidation as the result of proliferation and dissolution of plausibility structures.

In my view, this sort of consequential intermingling and rivalry among plausibility structures is the very essence of pluralism. Pluralism requires that we recognize the need for choices where fate once held sway. Thus, Berger insightfully expresses the view that for a church to survive in a pluralized world where choices are abundant and destiny has lost its meaning the church must take on the elective character of a voluntary association. A voluntary association invested with a sacred character, even while it embraces choice and eschews fate, is difficult to conceive. This would seem clearly to be an instance where the secular trumps the sacred. In my view, it may also render problematic, of uncertain meaning, Berger's oft-repeated claim that much of the world is becoming more religious, and secularization theory has been thoroughly falsified.

The concept "sacred" appears frequently throughout Berger's work, complementing rationality and perhaps competing with it for preeminence as a constituent of a world view (Berger [Ed.], 2009). Since he is a sociologist of knowledge with a life-long interest in religion, the conspicuousness of references to things sacred makes a good deal of intuitive sense. However, when I sit down and try to produce a definition for sacred, I find that I come up short, failing to produce anything that strikes me as satisfactory. Fortunately, early in *The Sacred Canopy* (Berger, 1967) Berger gives us a definition that I think is quite useful:

> By sacred is meant … a quality of mysterious and awesome power, other than man and yet related to him, which is believed to reside in certain objects of experience. This quality may be attributed to natural or artificial objects, or to the objectivations of human culture … Space and time may be assigned the same quality, as in sacred localities and sacred seasons. The quality may finally be embodied in sacred beings from highly localized spirits to the great cosmic divinities (Berger, 1967: 25).

Again, when nothing is eternally permanent lest it be inconsistent with functional rationality, and nothing is sacred lest it interfere with the creation and use of functionally rational means of production, some social outcomes are easy to anticipate: specifically, alienation, anomie, and egoism may quickly come to the fore (Marx, 1844/1988; Durkheim, 1897/1997). Man does not live by productivity, predictability, and profit alone, and to think and act otherwise is to invite intense resistance. Various forms of this resistance are sometimes referred to as demodernization or counter-modernization, a struggle for the stability of traditional meanings and values when faced with the onslaught of the ever-changing products of functional rationality (Berger & Kellner, 1981; Berger, 2011).

Styles of consciousness: the sociology of knowledge

We also see that modernization is not merely something that takes a variety of physical forms. It also takes shape in consciousness in ways that are sometimes salutary and sometimes destructive (Berger, Berger & Kellner, 1973). According to Berger, sociology and social theory have their genesis in an effort to understand the metamorphosis to what we now call modernization (Berger, 2014). Lived experience within the context provided by modern institutions has profound implications for the ways in which we think and feel. If this were not the case, rationality itself would be a concept attributable, perhaps, only to computers and robots. After all, we acquire the ability to be functionally rational only through participation in our material and social world, a set of circumstances that has become pervasively rational in its organization, more so as time goes by (Berger, 1963).

Discussions and concerns such as these bring to the fore the sub-discipline known as the sociology of knowledge. Berger may be its most eminent living practitioner, and he construes it as the sociological understanding of the determination of thought (Berger & Luckmann, 1966). A fundamental assumption of the sociology of knowledge is that thought is the product of lived experience in specific settings. The character of the settings, and our relationships to them, are the antecedents of what and how we think. They are the social sources of what we take to be real, and they pattern our responses to the world as we construe it (Berger, 1963). It is difficult – perhaps impossible! – to think of a sociological sub-discipline to which the sociology of knowledge would not usefully apply.

Reflection on the assumptions that undergird the sociology of knowledge may bring to mind the well-known and often dubiously employed concept of "ideology," a term we've already used several times. Given that what and how we think is determined by social circumstances, it's easy to imagine that the knowledge produced by everyday participation in a specific social setting may work to the advantage of some and the disadvantage of others.

For example, those who have spent their lives living and working in a distinctively rural setting with distinctively rural institutions and values may rightly feel effectively disenfranchised when their children's small schools are replaced by a much larger consolidated school (Howley, Bickel & McDonough, 1997). Frequently, this is because decision-makers representing state departments of education are urban born and bred, they have attended colleges and universities in urban areas or areas where urban practices and values prevail, and as professional educators have been imbued with the notion that large schools offer more opportunity. It's something they take for granted. Decision-makers' lived experience, thus, is antithetical to the knowledge and values of rural people (Bickel & Howley, 2002).

This is a set of circumstances where the experiences of a dominant group have suffused them with the substance and way of thinking of a particular faction, namely those with urban values that are sometimes at odds with rural standards (Schafft & Jackson, 2010). This, however, need not be consistent with higher achievement, nor need it be simply a self-serving effort at scholastic empire building. Instead, it reflects the ongoing urban socialization of local and state-level decision-makers. Insofar as it serves their particular social interests in this way, it may usefully be categorized as an ideology.

This example shows us that ideology, the content and style of thinking that benefits particular groups, need not be the product of willful duplicity or narrowly self-interested pursuit of payoffs. Given the life experience of our hypothetical group of urbanized decision-makers, it is the product of socialization. Thus, as Berger puts it:

> The task of the sociology of knowledge is not to be the debunking or uncovering of socially produced distortions, but the systematic study of the social condition of knowledge as such (Berger & Luckmann, 1966: 12)

Commonsense knowledge of everyday life

To make the social world navigable, our mundane interpretations, based on everyday experience and commonsense knowledge, may unself-consciously and provisionally grant indisputable permanence to things as they seem (Berger, 1963). Even though we may know better, it would be difficult to get by if we were not able to proceed in this way.

Wholesale failure of ordinary meanings in a world that we have come to participate in as taken-for-granted may give rise to the cultural deregulation and meaninglessness that Durkheim (1897/1997) referred to as anomie: painful, destructive, and becoming more pervasive in the modernized world that we currently share. This holds in spite of the fact that Berger sees an antidote to anomie in enhanced religious participation (Durkheim, 1897/1997; Horkheimer, 1947/2013). Social relativity, when we are forced to face it, may seem a good deal more discomfiting than liberating.

Perhaps this kind of experience is a powerful motivational source for radical Islam and other forms of zealotry, an experience that Berger has referred to in a suitably forbidding way as "anomic terror!" This is a brilliantly wrought concept of the kind that brings sociology to life and makes its concepts powerfully real (Berger, 1967). Avoiding meaninglessness is an increasingly difficult and profoundly threatening problem in our world (Krasko, 2004). This may very well be one reason why Berger is able to find organized religion advancing in breadth and intensity across much of our planet (Berger, 1992 & 2014).

Objectivity and error

Those who would take the foregoing discussion to mean that sociological knowledge is contingent and transitory have a point. Berger, however, offers a determined effort to make the case that sociology is in an advantageous position with regard to producing objective knowledge that cannot be dismissed as just relativistic speculation dressed up in methodologically complex and theoretically elaborate social scientific garb. That he sometimes steps away from this position, making fatal errors of his own, particularly his firm, affectionate, and unqualified embrace of capitalism, has already been acknowledged, and our attendant confusion has been made emphatically clear (cf. Berger, 2011).

For the most part, however, Berger's position with regard to objectivity is perfectly tenable and devoid of special pleading for the discipline of sociology.

Objectivity is manifest in empirically verifiable factual information, such as the median age at death in Slippery Rock, Pennsylvania for the year 2014. Any errors, such as inevitable failure to report the age at death of someone who was born, lived his life, and died alone and apart from everyone else will adversely affect the accuracy of the median, but errors such as this are not due to lack of objectivity. Besides, as a non-participant, it may be useful on theoretical grounds that isolates be excluded, perhaps to assure that our data is not subject to systematic bias.

Suppose, however, that given every effort to gather information objectively, we find that the median age at death is exactly ninety years and six months, much higher than the national median (Poston & Bouvier, 2010). At this point we continue to strive for objectivity, but we may also want to interpret this information, perhaps making some interesting, even useful sense of it.

If sociologists, demographers, or other social scientists have consistently found a positive relationship between median age at death and the frequency and quality of routine social interaction, we may endeavor to objectively measure these as they occur in Slippery Rock. Assuming that those who died in 2014 were from a population much like the one that now prevails, we may investigate this set of relationships as it manifests itself across the year in which we are living. Errors in measuring the median age at death, as well as the frequency and quality of social interaction, however we have defined them, will no doubt exist. However the errors are just that – errors – they do not compromise objectivity, especially if I am building on, perhaps replicating, very similar analyses done by others.

If I find a positive relationship between median age at death and my measures of beneficial social interaction patterns for residents of Slippery Rock, I may interpret it to mean that my findings are consistent with those reported by others: People who interact frequently in a wholesome and supportive way with friends and neighbors will, on average, live longer. Notice that as soon as I begin selecting factors that might account for or explain an unusually high median age at death in Slippery Rock, I begin to interpret, and I do so objectively.

If I had a very different outlook, however, I might respond to longevity in Slippery Rock with something like, "Ah, these must have been morally upright people and God smiled on them, granting them the gift of longevity." Again, I am interpreting, but now my sociologists' objectivity has been badly compromised by my unwillingness to step out of the relevance structure of a devoutly religious believer into that of a determinedly objective social scientist.

Too often, I think, we confuse measurement problems with lack of objectivity. As we can see, however, we may measure and observe with random error and still be objective. Objectivity inheres in our relevance structure, not in the approximate perfection of our tools.

This applies, moreover, not only to quantitative research, but to qualitative research, as well. Ethnography need be no more subjective than survey research or an experimental design. To the degree that the ethnographer adopts the relevance structure of a sociologist his work is objective. His observations may contain errors and his interpretations may be erroneous, but it's his relevance structure that determines the objectivity of his work, not his ethnographic or other sort of methodological skill. I have to admit that this is something I did not fully appreciate until I read *Sociology Reinterpreted* (Berger & Kellner, 1981).

This is a good place to acknowledge that I have found *Sociology Reinterpreted* (Berger & Kellner, 1981) to be a very useful book. It's quite brief, but it's also remarkably informative. It's repetitious, but this serves the useful purpose of functional redundancy. Some of its examples are not the best, but all told it's a book that merit's close attention. I'm not at all convinced that the authors have thoroughly reinterpreted sociology, but that's an awful lot to expect of any pair of social scientists in one brief book. As it is, they have written a readable and very instructive volume, an important part of Berger's total output.

Much more could be said about Berger's prose, presentation, and the purpose of his work that directly and indirectly highlights the overriding importance of institutions and institutionalization. However, I think we have seen enough to appreciate the interesting, difficult, insightful, and frustrating work of an eminent sociologist, one who specializes in the sociology of knowledge, and who has changed the way we view the world.

3 The sociological tourist

In my view the rigor and efficacy of "sociological tourism," both its strengths and its limitations, are inherent in the quality and pervasiveness of the work Berger has produced. A "sociological tourist," however, need not be stubbornly at odds with every kind of methodological standardization and uniformity, as illustrated by Berger's use of "cognitive maps." At least a modicum of methodological and conceptual flexibility is necessitated by the differing character of those being studied and by a variety of research objectives.

As one might expect of a sociologist or practitioner of any other mode of disciplined inquiry, Berger, throughout his large body of work, makes frequent reference to data, even using quaint, awkwardly worded, uncharacteristically positivistic-sounding statements such as "if the data fit"(Berger & Kellner, 1981), "when those theories are supported by the data" (Berger, 2011), and, "the most important thing I must do is search for those data that falsify rather than support my hypothesis" (Berger 1986). It may be that as an inveterate theorist of substantial influence Berger feels no obligation to repeatedly explain throughout his prolific output just what he means by data from one research project to another, or even whose data he's talking about, his own or someone else's in the unlikely event that he's doing a secondary analysis.

However, in view of his emphatically pronounced aversion to positivism, which he sometimes construes to mean anything quantitative, it would be useful if he were more forthcoming. In some instances, it seems clear that he is referring to observational data, the kind that is produced by an ethnographer. Berger's master's thesis and doctoral dissertation included participant-observation as well as historical data (Berger, 2011). This relative clarity of the meaning of data, however, is not typical of Berger's work.

It's good to know that data are, in some usually undefined sense, gathered through application of sociological concepts. It's also illuminating to know that the concepts must be vested with an awareness of the everyday meanings employed by those whose lives, relationships, and institutions are being studied (Berger & Kellner, 1981). But more information would be useful, enabling us, among other things, to better understand just how a sociological tourist does his work.

Adopting relevance structures

In *Sociology Reinterpreted* (Berger & Kellner, 1981), Berger provides a series of topically related hypothetical examples of how one might go about acquiring a better understanding of specific, sociologically informed aspects of a small part of the social world. As his illustrative efforts become more thoroughly interpretative, I take them to mean that he is making an increasingly determined effort to adopt the relevance structure of the hypothetical other with whom he is conversing, while at the same time maintaining adherence to the relevance structure of the sociologist. That is indeed an intellectually demanding twofold task.

I don't think the examples are particularly useful for one who might want to learn how to do sociology, but *Sociology Reinterpreted* is not a research methods text. Besides, Berger has an unfashionable view of the nature of useful research methods, one that works for him but might be too taxing or exceed the interpretative capabilities of most others.

More to the point, it seems reasonable to conclude from what Berger presents that adopting the relevance structure of another is a difficult but powerfully informative practice that has become second nature to him. The same seems true of his degree of professional detachment and awareness of the pertinent substance of his discipline needed to accomplish this effectively while simultaneously applying it to understanding the relevance structure of the other (Berger & Kellner, 1981).

This, I'm sure, is an important part of what it means to be a sociological tourist (Berger, 2011). While the process remains open-ended, non-formalized, and unstandardized, it's a good deal more demanding than the simple characterization "sociological tourist" suggests. As presented by Berger, moreover, when using the relevance structure of a sociologist, sociological tourism is thoroughly objective. Anyone who has done interpretation of a truly informative kind with regard to issues of significance knows that it is very difficult, and in truth the process never ends, the sociologist just moves on.

Still, the nature of what Berger calls "data" remains vague, though reasonable surmises are more or less easy to come by. It would be helpful, however, if he told us practical, seemingly inconsequential things, such as whether or not he takes field notes. As things stand, we don't even know why Berger sometimes embeds "data" in qualifying quotation marks and sometimes doesn't.

"Empirical testing" and how he does it raises still more difficult issues (Berger, 2011). By now, however, though Berger is the one who raised the topic, he has reached a point in his career where he can dismiss such annoyances with something oblique that diverts attention from him, something like "it depends on what kind of research you choose to do." Clearly, he is outside the realm of sociological tourism, in alien territory that he may commonly avoid when deriving concepts, collecting data, or thoughtfully formulating, rejecting, and reformulating hypotheses as they occur to him. In the very act of being a sociological tourist, Berger may be testing and re-testing hypotheses, but this is likely to be something he does very informally, something that goes on in his suitably well-stocked head.

Testable hypotheses

By contrast, in *The Capitalist Revolution* (Berger, 1986), Berger purports to present fifty testable hypotheses concerning capitalism. However, improper wording – unacceptably vague phrases such as "long lasting" and "ever expanding" – and presenting two or more propositions as if they were one renders the hypotheses inconsistent with any kind of formal testing. We can see what he's getting at, but we can also see his methodologically fatal errors. One wonders why Berger's editors were so conspicuous in their flagrant absence. Reworking Berger's propositions to make them testable would not have been particularly difficult, nor would it have required much formal methodological knowledge, little or nothing beyond a first course in research methods. On the other hand, unless he actually subjected his hypotheses to empirical test, a monumental and unlikely data-intensive undertaking, little would have been gained by presenting them in proper form.

I think it's reasonable to attribute Berger's difficulties in generating testable hypotheses to simple lack of practice and the absence of studied awareness of pertinent methodological details, even if they are easy to understand and execute. He has spent so much time doing fruitful sociological tourism and so little doing ostensibly more rigorous ways of practicing sociology that errors in the formulation of testable hypotheses are to be expected.

In any case, invoking the notion of data in a bland and taken-for-granted way, without further explanation as to what qualifies, seems off-the-mark and uninformative. It's as if Berger were trying too hard to make the case that he can be a hard-nosed social scientist, not just a casual observer who reads a lot of books, but his effort is unconvincing.

Perhaps Berger is not persuasive in this regard because he also revels in the role of a methodological maverick. He wants to have it both ways: disciplined social scientist and sociologically gifted gadfly who rejects the crudely positivistic tools that pass for "rigorous" in his discipline.

It's true, however, that in Berger's discussion of "evidence" in *Sociology Reinterpreted* (Berger & Kellner, 1981), he takes the position that both qualitative and quantitative data are admissible. The decisive factor, again, being that the subjects' own typifications and meanings are manifest in data collection and interpretation, still a difficult and effort-intensive task. Given the kind of information he may have collected as a sociological tourist, empirical testing seems too formal to apply. We've already acknowledged the difficulty Berger has in producing formally testable hypotheses.

Nevertheless, Berger's judgment that quantitative methods have become an unfortunate fetish in contemporary American sociology is emphatically true, and it's welcome to hear this assessment from a sociologist of his fame and attainments, especially since his own approach to doing sociology is far out of the discipline's mainstream. It's also interesting to find in Berger's earlier work, especially *Invitation to Sociology*, judgments that make clear that he did not foresee the current obsession with increasingly arcane, difficult, and sometimes dubious quantitative methods. The following quotation is instructive:

> Since about the mid-1940s there has been a revival of interest in sociological theory, and there are good indications that this tendency away from narrow empiricism is continuing to gather momentum (Berger, 1963: 9).

I suspect that Berger was referring to the influence of theorists such as Talcott Parsons, Robert Merton, George C. Homans, C. Wright Mills, Lewis Coser, and Peter Blau, though there are certainly many others. Whoever he had in mind, the movement away from narrow empiricism has long since lost momentum. Parsonian scholasticism lasted a bit longer, but it too, in its staid and effusively bombastic prolixity soon lost its once ironclad hold on what passed for social theory (Berger, 2014).

In short order, in fact, any respectable sociological journal that published an article not including a path model with more or less interpretable path coefficients and explanations as to why some paths had been dropped, was no longer respectable. This was a set of circumstances lamented by Lewis Coser when he was President of the American Sociological Association (Coser, 1975). Sociologists' quest for acceptance as scientists may have been the motive force that gave rise to a proliferation of new and different statistical techniques whose dominance shows no sign of abating.

In view of Berger's lifelong inclination toward labor-intensive methodological informality, he could afford to ignore most of this. Had his career as a sociological eccentric been less successful, he might have been forced to modify his posture. It's to his credit and our benefit that this did not happen. Historically, philosophically, and linguistically informed, rather than sterilely schooled in the often conflicting methods of contemporary empiricism, Berger has made numerous favorably influential – sometimes brilliant and lasting – contributions to the theory and substance of sociological knowledge.

Mainstream sociology, as Berger understands it, has devolved into what he terms the "twin distortions" of ideology and triviality. As a result, the really important questions that sociologists should address are ignored, while those addressed are increasingly limited to ones that lend themselves to the status of empty exercises in methodological retrenchment and self-interested forays into questions that are high-sounding but of little or no interest (Berger, 2014).

"Cognitive maps"

It's a bit out of character for Berger that, while a member of the previously mentioned disparate group of scholars and activists examining the possible ways in which South Africa might divest itself of apartheid and what it would look like afterward, he made a methodological development in which he seems to take enormous pride. It represents, moreover, a sharp departure from sociological tourism, something I had not expected (Berger, 2011).

Berger recognized that there was consequential diversity in this group, just as there was a great deal of variability among stakeholders throughout South Africa. As a result, using his concept of "cognitive maps," he directed members of the group to participate by producing their own maps, including routes to freedom from apartheid so starkly different that some envisioned a violent uprising while others relied exclusively on moderation and incrementalism. Once Berger had a manageable set of cognitive maps, capturing the social character and procedure intrinsic to each, he circulated them, asking just one question: Is the strategy likely to succeed?

According to Berger, the method was very effective, though he doesn't make clear what effectiveness or success mean in this context, nor are we certain as to the nature of his data. Nothing but answers to yes/no questions? Could be, but still, he had enabled a group of shareholders to participate, even if at his direction and according to his standards. The upshot was that Berger was so pleased with the results of application of his method that he continued using it in a variety of other projects.

The main reason I have introduced this vaguely defined method is because, to my knowledge, it represents one of the few instances in which Berger explicitly acknowledges taking leave of sociological tourism. Whatever the precise nature and merits of his new method, it was roughly standardized and crudely formal. As alien to most of his work as the characteristics standardization and formality are, it still pleased Berger very much, casting him in the role that C. Wright Mills recommended in *The Sociological Imagination* when he told each sociologist to be his own methodologist, certainly much easier said than done, but Mills was not known for exercising caution and restraint (Mills, 1959).

Otherwise, Berger seems quite content to continue to characterize his usual methodology as determinedly informal sociological tourism. This brings to mind the longshoreman-philosopher Eric Hoffer's description of himself as one who "went through life like a tourist" (Hoffer, 2006), though he still managed to write brilliant books. Not an unflattering comparison for either Berger or Hoffer.

I admit that it is refreshing to find an accomplished sociologist such as Berger not belaboring usual methodological concerns but acknowledging what I take to be a claim that his work is due to omnivorous reading, both casual and very careful observation, and thought. It is also the case, however, that I find Berger's peculiar methodological approach less appealing when I disagree with him and question the merit of his judgments.

Nevertheless, and perhaps I'm asking too much and being decidedly wrongheaded, Berger's explanation of data, skimpy or not, is missing from most of the work in which he uses the term. In this regard and this regard only, Berger reminds me of Talcott Parsons, someone whose efforts might be taken more seriously even today if his ponderous methodological pronouncements had been written more clearly and had been less anachronistically stilted and better informed by tangible research experience (Parsons, 1951/2012). Berger's work is better than this, in large measure, I suspect, because he has actually done a substantial amount of research himself, but he still leaves predictable and important questions unanswered (Berger, 1986).

Berger's research in social context

I don't doubt that Berger knows very well from firsthand involvement that research is a messy, uncertain, frustrating endeavor. Again, however, I do wish that he gave us more readily accessible and purposefully applicable information about how he actually does it. Truth be told, as already suggested, I'm guessing a good deal as to the precise nature of sociological tourism. Berger's hypothetical examples in *Reinterpreting Sociology* (Berger & Kellner, 1981) are contrived, exaggerated and otherwise unrealistic. As a result, they are not nearly as instructive as they might be, especially since they are included in such a readable and interesting book.

Sadly, in the United States, research methods have come to be construed very narrowly, limited to choices regarding how best to collect and analyze empirical data. The data may be quantitative or qualitative, with modes of analysis tailored accordingly. Moreover, even qualitative data may be analyzed using canned computer programs (see, for example, Friese, 2012).

Tools like this have been available since content analysis was computerized. I think, however, that they rob ethnography – and would rob sociological tourism – of the real power of its interpretative character and the originality of its insights.

It should come as no surprise that the more complex the statistical techniques applied to quantitative data, the more formidable the research is taken to be. As Berger recognizes, methods of quantitative analysis have become a fetish or obsession in American sociology.

Methodology, however, need not be construed so narrowly and taken to such positivistic extremes. Instead, it may refer to a set of assumptions and a point of view concerning how best to think about and make sense of the social world in which we are located and the nature of our peculiar position in it (Gadamer, 1960/2004). In that singular conceptual sense, Schutz's, phenomenology can be understood as a method (Schutz, 1967). Schutz's influence on Berger is difficult to exaggerate, providing the phenomenological relevance structure for much of the sociology that Berger has done.

Using Schutz's relevance structure as a conceptual tool, Berger argues that sociologists have the training to participate in a richly informed tradition that provides them with a favorable vantage point, a way of selectively perceiving and making sense of the infinite variety of stimuli with which they are inescapably confronted. This powerful process of secondary socialization or internalization may be transformed into a social scientific occupation, with its unique point of view focusing the sociologist's attention and interpretative interests using a mindset and concepts that help to filter out the manifold distortions intrinsic to any social situation (Berger, 1963; Berger & Kellner, 1981; Berger & Huntington, 2002).

The sociologist, in effect, has learned to discount, for the time being, the relevance structures that he uses in getting through the day as an ordinary citizen (cf. Garfinkel, 1967). The concepts and interpretative procedures used by the working sociologist are thus restricted to those intrinsic to sociology as an established discipline, and all the rest is pushed aside until the sociologist returns to the role of an ordinary person. Thus Berger's use of the term "dual citizenship," another conceptual and linguistic innovation that serves him well and succinctly clarifies an important aspect of the sociologist's role (Berger, 2011).

The work of the middle class sociologist, as a result, need not be subverted by the fact the he may be studying peculiarly middle class issues that he encounters every day. He avoids this epistemological trap by adopting the relevance structure of the professional sociologist, tossing aside for the time being his conventional middle class perspective. It is by adopting the relevance structure that characterizes his discipline that the sociologist approximates objectivity (Berger [Ed.], 1990).

4 Ever the Weberian

While Berger is best known as a humanistic sociologist who appreciates sociology's close ties to history, philosophy, and literature, having read his earliest books decades ago I was then inclined to characterize him as a philosophical materialist. By that I meant one who, when all is said and done, gives priority to lived experience or material practice in the acquisition of the themes, symbols, and ideas that constitute his unself-consciously held world view and define his internalized social role. Hence the priority of institutionalization.

Presented this simply, however, Berger might very well strenuously object, detecting, perhaps, an unduly close similarity to the perspective of, say, an excessively deterministic Marxist (Althusser & Balibar, 1970), as well as an undeveloped rendering of lived experience (Berger, 1963; Berger, Berger & Kellner, 1973). In any case, I know of no instance in which Berger has explicitly expressed a disciplinary kinship with philosophical materialism.

Berger is emphatically clear that the classical sociologist with whom he has the most intimate affinity is Max Weber. True to his early even-handed, value-neutral approach to social theory, however, Berger also acknowledges the contributions, along with what he takes, sometimes sardonically, to be the limitations, of Marxist perspectives on contemporary social phenomena (see, especially, Berger & Luckmann, 1966; Berger, 1967; Berger, 1986; Berger [Ed.], 1990). To his credit, Berger acknowledges that the sociology of knowledge took its most basic proposition, that "man's consciousness is determined by his social being" from Marx. Berger also expresses approval of the contribution of Durkheim, though he is not fond of Durkheim's positivism (Berger & Kellner, 1981; Berger, 2014).

Berger on Marx and Durkheim

I think, however, that Berger oversimplifies claims that Marx is a Utopian, heir to all the theoretical and methodological difficulties intrinsic to that status. Moreover, he exaggerates the limitations that come with Durkheim's positivism. In truth, Berger seems to take Durkheim's self-stated positivism much too seriously. Unalloyed positivists don't contrive brilliantly insightful substantive macro-level concepts such as mechanical solidarity and organic solidarity. If Durkheim misrepresented the genuinely introspective and intuitive nature of his work, it wouldn't be the first time a social theorist, going back to Comte, has failed to see the real character of his own social thought (cf. Fournier, 2012).

Furthermore, Utopian or not, Marx's thorough understanding and effective use of his concept of alienation, in all the intuitively appealing forms he gives it, makes clear that he was sensitive to the lived experience of members of the class most important to him, the proletariat. In view of his close association with Friedrich Engels, moreover, it seems certain that Marx was influenced by *The Condition of the Working Class in England* (Engels, 1845/2009), and with its author's appreciation of the appalling, life-sapping character of existence as it was endured by English workers during the first half of the nineteenth century (cf. Himmelfarb, 1984).

Similarly, Durkheim's introduction and creatively innovative employment of macro-level concepts such as anomie and egoism are emphatically manifestations of understanding of aggregated subjective meanings that easily match in explanatory power and flexibility anything that Weber produced. Durkheim's continuing influence on contemporary sociology, in my view, is profound but largely unrecognized and taken for granted. It provides the substratum for much of contemporary sociological discourse.

In truth, I think that Durkheim's influence far surpasses that of Weber. Both Marx and Durkheim, however they characterized themselves and whatever their disclaimers, were capable students of the meanings that people applied to their circumstances and lives (Bickel, 2013).

Ideology and interpretation: Marx

Nevertheless, Berger is certainly correct when he asserts that purposefully adopting a rigid ideology, for whatever reason, closes off much of the world and distorts interpretation of participants' meanings and the structures in which they occur. Ideologies, in effect, prejudge important issues before adequately sustained efforts at interpretation have been made (Berger, 1963).

Moreover, an ideology brings to bear its own relevance structure, making alternatives extraneous and inaccessible. Oddly enough, Berger seems to have placed himself in just this position when, in his later work, he concludes that alternatives to capitalist efforts at modernization are endeavors he could not evaluate evenhandedly because of his strong commitment to capitalism (Berger, Berger & Kellner, 1973; Berger, 1986; Berger [Ed.], 1990; Berger, 2011).

I do not understand how Berger justifies his unquestioning devotion to one way of accomplishing the very complex and only imperfectly understood processes that generate and maintain modernization, especially when he acknowledges that modernization can be approached in a variety of different ways (Berger 2002 & 2014). It seems a betrayal of sociology as Berger understood it in his earlier work (Berger, 1963; Berger & Luckmann, 1966; Berger, 1967; Berger & Kellner, 1981).

I do think, moreover, that while Berger gives voice to respect for substantial research and theorizing based on a Marxist perspective, he is too quick to conclude that proponents of Marxism close off interpretations and deny the value of relevance structures not based on Marxist themes, concepts, and presuppositions. In my experience, most people who have acknowledged a significant indebtedness to Marx recognize that his work, as with other social scientific perspectives, has the incomplete and partial character that Berger applies to the social sciences generally (Horkheimer & Adorno, 1947/2007; Thompson, 1963/1980; Gouldner, 1980; Eagleton, 2012).

Furthermore, Berger's (1986) condemnation of Marx's work for introducing philosophical assumptions in the form of his labor theory of value, in place of social scientific concepts that can be organized into testable propositions, has implications that seem to elude him. After all, Adam Smith, David Ricardo, Jean-Baptiste Say, and Thomas Malthus all subscribed to one form or another of the labor theory of value, as did most other nineteenth-century economists (Whitaker, 1990).

Marx's take on the labor theory of value was not so distinctive that it rendered his economics hopelessly flawed, while others remained scientifically viable. Furthermore, contrary to Berger (1986) there is nothing in Marx's theory that is in principle unmeasurable (Sweezy, 1970; Mandel, 1994). Units of analysis for some concepts may be very difficult to construe and as a practical matter unworkable. However, one might very well take that to mean that there is more research to be done, as Marx would have been the first to acknowledge.

Nevertheless, given the radical and influential nature of Marx's overall perspective, his labor theory has been subjected to more intense scrutiny than the labor theories of others. John Maynard Keynes's student and colleague Joan Robinson wrote a caustic and influential essay in which she rejected Marx's labor theory in favor of an alternative she developed directly from it but applying ideas taken from Keynesian economics (Robinson, 1942/1967). The Norwegian-American political economist Thorstein Veblen rejected all labor theories of value, Marx's included, as unduly metaphysical (Veblen, 1904/2013).

I juxtapose authors as different in time and substance as Robinson and Veblen because of their scholarly stature and influence, and because both were fundamentally sympathetic to Marx's overall theoretical perspective. If Berger were consistently the even-handed scholar he succeeded in being in his early work, he would not have given a short-shrift, summarily dismissive account of a central concept of the scholar he identified as his principal antagonist in *The Capitalist Revolution* (Berger, 1986).

After all, Berger (1986) chose Marx's labor theory of value as his point of critical entry. I think this obliges him to give it creditable treatment. Four or five paragraphs would have been enough to put Marx's work into a suitably informative, even if very abstract, historical perspective, and to avoid simple sloppiness in giving voice to its primary deficiencies.

In truth, there remain accomplished contemporary scholars who have worked honestly, skillfully, with substantial learning and technical acumen to better understand Marx's labor theory of value (Mandel, 1994; Martin, 2008; Harvey, 2010; Sowell, 2014). Marx's effort is much more than the social scientific joke to which Berger (1986) seems inclined to reduce it.

Yes, there are dogmatic adherents who stupidly regard the Marxist texts as genuinely sacred. Even among them, however, there can be contentious differences as to what Marx really meant, such as whether or not the early and later texts are compatible, and whether or not the base/superstructure model, as homologous with usual understanding of plausibility structures (Berger, 1986), is of theoretical interest (Bottomore, 1956: Martin, 2008). Even a true believer such as Poulantzas (1978/2015), given his original and fertile mind, was open to and adopted modifications in some of Marx's centrally important concepts.

Furthermore, anyone who has read Raymond Williams's *Marxism and Literature* (1978) or Terry Eagleton's *Criticism and Ideology* (1976) will quickly see that when Marxism ranges broadly, say into the realm of aesthetics, the concepts used are quite different from more conventional Marxist analyses. This holds even though the endeavors, sometimes with good reason and sometimes without, are still characterized as Marxist.

Unfortunately, despite the openness and theoretical flexibility that characterized Berger's early work, especially *Invitation to Sociology* (Berger, 1963), *The Social Construction of Reality* (Berger & Luckmann, 1966), and *The Sacred Canopy* (Berger, 1967), by the time he got to *The Capitalist Revolution* (Berger, 1986), his endeavors aimed at explaining modernization, economic development, and the social concomitants of capitalism had become unduly contentious and, paradoxically, bland. Perhaps one reason that Berger's ostensibly testable hypotheses were ineptly formulated is that Berger had already made up his mind and didn't pay much attention to crucial details in what he was doing.

Berger starts off by characterizing Marx as "one of the intellectual giants of the modern age." From then on, however, his treatment of Marxism is too often ham-handed, misinformed, and uninformed. The dearth of credible recent references – or any references at all – is inexcusable in a book such as *The Capitalist Revolution* (Berger, 1986) that, in large part, can be fairly interpreted as an anti-Marxist polemic (Berger, 1986).

In writing *The Capitalist Revolution* (Berger, 1986), Berger seems not to have noticed that the uniformly favorable good economic times of the post-World War II era had already begun to fade rather badly. As we approach the present day, some of Marx's forecasts that Berger dismissed as contemptibly obvious nonsense have gained a good deal of plausibility (Eagleton, 2012).

I don't know if the fact that in the United States, working class has become coterminous with an impoverished social location, while the wealthiest absorb almost *all* the aggregate gains in income and wealth, is a rough approximation of Marx's concept immiseration (Marx, 1867/2011). Whether or not it is, it's a manifestation of widespread misfortune and inequity that Berger did not foresee, and it is sharply inconsistent with his claims that levels of inequality and upward mobility would remain approximately stable from the 1950s on, with "on" meaning through the present and into the indefinite future (Berger, 1963 & 1986).

This is quite different from the equity versus income issue, with all classes having enough to get by but some having appreciably more. The problem now is that many people are painfully poor in an enervating, disease-producing, shortened-life-span fashion. Many are homeless in the brutally roofless physical sense (Liebow, 1995).

Others, moreover, while still identifiably middle class, are threatened with falling into oppressive poverty, while members of a relatively small, identifiable upper class are consistently receiving enormous payoffs (Domhoff, 2013). (We will address the preposterous, mean-spirited claim, popularized during the 1970s, that poverty is inherently relative later in our discussion.)

Given the foregoing, if an interlocutor contended that this was, indeed, a portent of immiseration in a thoroughly modernized capitalist country, Berger, I'm certain, would dismiss the contention as nothing more than the refusal of vulgar Marxists to surrender when it was abundantly clear to all that they had been defeated (Berger, 1986; Bickel, 2013). Maybe so, but in the world of late capitalism things change quickly (Crary, 2014).

Ideology and interpretation: Durkheim

In my experience, Durkheim can be an especially important source of clarifications, extensions, and straightforward modifications in Marxist thought, though Durkheim's value in this regard is rarely discerned or acknowledged. In good part, this oversight may be due to Durkheim's emphatically self-professed, perhaps unduly exaggerated positivism.

Though positivism rejects introspective and intuitive knowledge, as noted above, it's difficult to imagine how Durkheim could have derived some of his most influential and fruitful concepts, especially macro-level ideas pertaining to his research on suicide (Durkheim, 1897/1997), without resorting to prohibited procedures. Anomie, egoism, mechanical solidarity, and organic solidarity, especially, can be useful as macro-level explanatory factors in analyses prompted by capitalist development as Marx expected it to occur. I found Durkheim's concepts especially useful as theoretical supplements in quantitative analyses of elementary school achievement based on a fundamentally Marxist point of view (Bickel, Smith & Eagle, 2002; Bickel, Howley & Maynard, 2003).

Analyses of school achievement and other school-related outcomes could readily be construed as analyses of one especially important aspect of our "economic culture," a term that Berger uses to refer to non-economic implications and manifestations of capitalism (Berger, 1986). In this instance, the nature of the non-economic outcome is public schooling, long viewed as an agency of upward social mobility, though extravagant expectations for investments in secondary and even higher education have recently been sternly tempered by disappointing outcomes (Perkinson, 1968/1995; Putnam, 2000). In truth, economic culture as used by Berger refers to "the interface of economic institutions and various elements of culture [such as ideas, religion, morality, and lifestyles]," a much broader scheme than Berger later offers for capitalism itself, that "lends itself beautifully to elaborations in terms of the sociology of knowledge" (Berger, 2014: 13).

Most of the research that I did over the course of thirty odd years as a program evaluator, policy analyst, and academic was quantitative in character, and much of it was heavily indebted to Durkheim. However, it would be a gross misrepresentation of this material to suggest that it was not interpretative and therefore at odds with what is peculiarly human.

As with applications of Durkheim's concepts as explanatory factors in analyses based on a Marxist perspective, the concepts I used most frequently, operationalized more or less adequately, were anomie or cultural deregulation, egoism or deracinated homelessness, and organic solidarity or departures from traditional values and practices. I used these concepts, among others, to try to make sense of social phenomena such as rates of teen pregnancy (Bickel, Weaver, Lange & Williams, 1997), rates of early teen pregnancy (Bickel, McDonough & Williams, 1999), rates of dropping out of school (Bickel & Lange, 1995), rates of reckless behavior among high school students (Bickel & McDonough, 1997), rates of crime on school property (Bickel & Dufrene, 2001), and degree of success in implementation of the centerpiece of George W. Bush's domestic policy, No Child Left Behind (Bickel & Howley, 2003), specifically in the Appalachian region of the United States.

Furthermore, some of the other resources I relied on most heavily were ethnographic, especially Paul Willis's classic, *Learning to Labor* (Willis, 1981) and Michael Burawoy's *Manufacturing Consent* (Burawoy, 1992). Used in conjunction with Durkheim's ideas, I found Willis's and Burawoy's ethnographic material to be invaluable interpretative resources, and Willis and Burawoy are definitely not positivists.

My point is that while Berger endorses use of quantitative research when he thinks it's employment is fitting, he has a tendency to conflate quantitative with positivistic, thereby denying the interpretative nature of much suitably executed quantitative research. Yes, we all know that quantitative methods of the most sophisticated sort have become an obsession in contemporary American sociology, and we have already acknowledged as much. However, quantitative and positivistic are not coterminous. Interpretation is an essential part of good quantitative research, assuring that Berger's interest in maintaining the human quality is not ignored.

Ideology and interpretation: technological determinism

In the not too distant past, it was not unusual to find some quantitative researchers derisively characterized as "multiple regression Marxists." As best I can determine, the scholar whose work did the most to elicit this label was Erik Olin Wright, with books like *Class Structure and Income Determination*

(Wright, 1979). Until Wright and a few others came along, Marxists who used sophisticated quantitative methods were limited to a handful of out-of-the-mainstream economists, including Valerie Nelson, and Samuel Bowles and Herbert Gintis, all of whom contributed to the social science of education classic *Schooling in Capitalist America* (Bowles & Gintis, 1975).

Before this eccentric band of researchers began to make its mark, quantitative research in Marxist studies was almost never done, and was typically dismissed by most other Marxists as just so much misguided and empty empiricism. Now things have changed.

Today we're more likely to encounter unwavering and devout technological determinists who brook no departures from commitment to the idea that technology is the source of all social change of any consequence, and in some extreme instances argue that human beings may be well on the way to obsolescence (Smith & Marx, 1994). As a result, the attention that Berger gives to Marxist ideologues seems excessive and inconsistent with the theoretical and methodological flexibility one often finds among those heavily influenced by Marx and his sometimes prolific followers.

After all, most Marxists are not active revolutionaries whose interpretations are in accord with an unduly zealous commitment. Furthermore, before the collapse of the Iron Curtain, most sociologists on the outside of that border of ideological constraint were not favorably disposed to what was passed off as sociological insight in the Soviet Union, though the Gorbachev Era occasioned increasingly interesting social thought.

In my view, the pervasive technocratic mindset is of far greater danger to the interpretative power of good sociology than Marxism (see, for example, Bell, 1972 & 1976; Zizek, 2011). Berger pretty clearly acknowledges as much at various places throughout his estimable body of work, much as we might expect of a sociological tourist (see, for example Berger & Kellner, 1981: 146–178.)

Technocratic social engineering may seem like a term so strong and certain in its implied efficacy as to be misleading when applied in a so-called soft science such as sociology. However, even when sociology is reduced to poorly conceived, lightweight journalism, technocratic manipulation can have painfully adverse consequences and also discredit sociology as a discipline worthy of serious attention.

If a journalist of dubious integrity, or for that matter an highbrow policy wonk, claims that sociologists have established the norms of adolescent sexual behavior, adolescents, as well as their parents, will take notice (see, for example, Crooks & Baur, 2013). Whatever the quality of the information, adolescents may take the expressed or implied norms as scientifically validated prescriptions for their own behavior. How are they to know otherwise?

Chaste adolescents may be shamed into becoming sexually active. Promiscuous adolescents may conclude that since they are exceeding the norms, they should continue on, beating the benchmarks, so to speak (Crooks & Baur, 2013). Sex education instructors may be influenced in what they teach, even though the sociology they've read is of abysmally poor quality, superficially popularized, and misleadingly taken out of the context of much larger and far more informative research projects. Thus, simply because sociology exists and some people take it seriously, self-fulfilling prophecies are generated.

As another hypothetical example, imagine that an article in a popular magazine purports to provide information from recent sociological research on income attainment. According to the magazine's hypothetical account, the median annual income of men who occupy the social role of head of a household is $50,000. Perhaps a man who is living with his traditional family quite comfortably, at least by their standards, but making only $40,000 reads the article and feels painfully diminished, no longer the capable breadwinner he thought himself to be.

Thus, we see establishment of a standard devoid of context that may in and of itself be absolutely without value or interest, but that has been used in a way that produces needless discomfort. Perhaps by offhandedly invoking rudiments of the language of sociological role theory the assertion was made still more compelling. Victimization by off-the-shelf sociology may then be made even more damaging, generating dissatisfaction where previously there had been contentment.

Moreover, for the alert observer who recognizes that popularized sociology is likely to generate self-fulfilling prophecies and to establish meaningless but perhaps damaging standards, the prospect may be unsettling. He may find it disquieting and dishonest that choices will be made in a determined way and emotions will be pointlessly and painfully manipulated, whether intentionally or not, under the guise of sociological research. More generally, insofar as he sees the world of journalism and electronic mass communication as replete with messages that distort free choice and honest feelings, he may find the whole thing, sociology and all the rest, simply repugnant.

These are hypothetical examples, but as with any presumably authoritative discipline practiced by experts, sociology can be misused, even if the misuse is more accidental, careless, or simply lacking in rigor rather than intentionally mean-spirited or self-serving. Berger follows examples such as these with the observation below:

> This "imperialism" of technocracy, both in the mind and in social praxis, is one of the serious contradictions of modernity. Inevitably there are resistances against this. These resistances, which in their final intentions are not just anti-technocratic but counter-modernizing have been present since the onset of the modern era. They are powerful realities in societies now just undergoing modernization (Berger & Kellner, 1981: 133).

It seems an overstatement to attribute this much social power to sociology, and it probably is. Nevertheless, if we pay attention, we can see every day that sociology, the other social sciences, as well as undisguised publish relations showboating and advertising are turned to the disadvantage of unsuspecting ordinary people.

As with any mode of discourse, sociology has its own relevance structure. When sociology is forced out of it relevance structure and an alien relevance structure subsumes sociological practice, what remains of its efficacy is likely to discredit sociology and serve the narrow interests of merchants, mercenaries, and others in pursuit of measurable, pragmatic, dollar-valued outcomes (Berger & Kellner, 1981). Berger understands this and it's to his credit that he has brought it to our attention.

5 Sociological insights into the exemplary, the everyday, and the execrable

Modernity, however, is another matter. Throughout his work Berger struggles ineffectively to provide a more or less satisfactory definition of modernity. This is especially disappointing, because the concept of modernity is central to so much that Berger has written. I suppose it's possible that modernity could be construed as an idea that does not require a formal definition because it is so familiar that it speaks for itself. In truth, however, this suggestion seems hopelessly ill-suited in the context of technical sociology done by one who has devoted a substantial part of a long career to clarifying the import and implications of varying levels of modernity on a world scale.

I suspect that I devote too much attention to this difficulty, as in the following section titled "Modernity Defined." However, an idea as crucially important as modernity deserves at least minimally adequate attention, something I try to provide.

In view of the fact that modernization and modernity have become preoccupations with Berger in much of his writing and research, *The Homeless Mind* (Berger, Berger & Kellner, 1973) provides useful, even if only occasional thematizing instances for various parts of our ongoing discussion. It effectively portends some of Berger's most compelling interests until the present.

Berger had not yet made his surprisingly strong and, I think, sociologically execrable commitment to capitalism, but in other respects *The Homeless Mind* (Berger, Berger & Kellner, 1973) is more or less typical of Berger's work on sometimes contentious instances of social change in modernizing and modernized environments. It also enables us to discuss important instances of Berger's productively creative contributions to social theory. We shall use *The Homeless Mind* (Berger, Berger & Kellner, 1973) episodically, coming back to it only as clarity permits and theory and substance make it useful. This is not a book review, but I'm using *The Homeless Mind* as a sort of mnemonic device regarding modernization and modernity (Berger, Berger & Kellner, 1973).

In addition, use of *A Far Glory* (1992) in much the same way serves the sacred and the secular, two more of Berger's favorite conceptual interests, by providing rudimentary scaffolding for the recurrence of those ideas. *A Far Glory*, too, alerts us to the presence and relative importance of what is to follow as we read through the body of Berger's material.

Furthermore, just as there is more of interest than modernization and modernity in *The Homeless Mind* (Berger, Berger & Kellner, 1973), there is a great deal in *A Far Glory* (Berger, 1992) that is not captured by the concepts sacred and secular. It is useful, therefore, to acknowledge the way our presentation develops, without being heavy-handed or massively inclusive, expecting too much from key words.

There is a truly peculiar irony, however, in the fact that *The Homeless Mind* (Berger, Berger & Kellner, 1973) was written as an application of the perspective developed in *The Social Construction of Reality* (Berger & Luckmann, 1966) to modernization and modernity (Berger, 2011). Though Berger may not have anticipated application of this brilliantly developed and clearly written point of view to the kind of material included in *The Homeless Mind* (Berger, Berger & Kellner, 1973), he had already played a principal role in producing a new and different perspective, employing original concepts of broad pertinence. Why he felt the need to go clumsily beyond that with ineffective, ill-favored neologisms simply escapes me, but I appreciate the irony.

Modernity defined

In the first chapter of the manuscript, we devoted a good deal of attention to the stylistically distinctive but sometimes substantively misleading fashion in which Berger presents his material, noting along the way that even brilliant scholars err. A primary rationale for the critical account was that inconsistent or illogical treatments of material of substantial importance were juxtaposed and concentrated in one book, *Adventures of an Accidental Sociologist* (Berger, 2011). I speculated that Berger's remarkable sociological achievements and imposing reputation may have led him to become less cautious than was prudent; in short, he may have been a victim of his own success.

In the present instance, instead of being prompted to make critical observations by the problematic contents of one book, the difficulty with which we are concerned covers much of the broad span of material that Berger has produced over the course of what is already a long career. In doing so, I raise the possibility that Berger may have long been boldly careless to his readers' disadvantage. My main concern is his mystifying and persistent failure to adequately define modernity, a concept of central importance in much that he has written.

Some readers may judge the meaning of modernity to be self-explanatory and not in need of definition. That this is not the case is abundantly evident throughout Berger's work. As he uses modernity in enormously varied contexts that defy amalgamation, we see that it is not only centrally important but quite complex. Left undefined, one wonders if Berger's invocation of modernity is sometimes just carelessly convenient, even slipshod. It could be that Berger, perhaps unself-consciously, sometimes unfortunately uses modernity as a buzz word.

If it had been stylistically, substantively, and topically workable to situate this example in Chapter 1 with the other troublesome passages that I cited, I would have done so. Instead, I have dealt with it at the point that it became troublesome to me as an engaged reader in need of a useful definition of modernity as Berger understands it and that illuminates his presentation.

As it turns out, I got a definition – in fact several, some of them oddly novel, some tediously mundane, some simply bad – but I'm not at all sure that I got a useful one. I thought I had found what I needed in Berger's recent book, *Pillars of Modernity* (Berger, 2014), but the definition quickly fell apart as Berger seemed to be saying that modernity and modernization are the same thing, a claim he would never knowingly make (Berger, Berger & Kellner, 1973).

Early in *The Homeless Mind* (Berger, Berger & Kellner, 1973), Berger apprises us of his aversion to using the concept modernity in a way that suggests progress. As he sees it, and I'm sure correctly, progress is a vague, unmeasurable, and value-laden idea that has no place as a technical concept in a sociological disquisition.

Though we may conventionally and commonsensically think otherwise, it's likely that even finding agreement as to what constitutes progress among the members of a specific population would turn out to be an exercise in futility. Furthermore, value judgments such as those embodied in the notion of progress are not within the legitimate purview of the sociologist, as Berger understands things. Using the thoroughly informative theoretical, substantive, and methodological material that constitutes a sociologist's professional background, he must adopt a point of view or relevance structure that pushes value judgments aside in favor of information that can reasonably be seen as objective and value-free, or so Berger has repeatedly argued (Berger, 1963; 1986 & 2011).

Thus, following Berger, a primary virtue of the concept modernity is that, in spite of the fact that it is frequently misused and remains definitionally elusive, he construes it as just such a value-neutral concept. According to Berger, modernity is an historical idea that should be regarded as such: a social scientific conceptual tool that betokens a beginning and end, and that can be studied in just that way. With well-informed hard work and a little luck, the social processes that give rise to modernity, promote its further development, and lead to its eventual demise can be identified and dispassionately studied.

For now, we will tentatively define modernity as a distinct consequence of the process of modernization made manifest in a specific set of social institutions and corresponding cognitive styles. As we move through the material that follows, I trust that the value of this very brief and very abstract definition will become evident. I'm sure that it's immediately clear, however, that ease of operationalization is not one of its favorable characteristics.

Many meanings of modernity in bits and pieces

Whatever the merits and deficiencies of our provisional definition of modernity, Berger keeps us guessing and forces us to make tentative inferences based on what might reasonably be described as the vaguest of offhanded hints as to his own definition. This is, at best, annoying; even more so, when he offers a hopelessly inadequate definition, or perhaps it's just an offhanded comment that I misconstrued, such as "Modernity means life with large numbers of strangers" (Berger, 1986: 32). This may very well be on the mark, bringing to mind Tönnies Gemeinschaft/Gesellschaft dichotomy (Tönnies, 1887/2001), but it gives us only one, seemingly superficial characteristic of modernity with no discussion of its meaning or consequences. Surely there's a good deal more to modernity than "life with large numbers of strangers" (Berger, 1986: 32).

This is the one and only time that Berger reminds me of Michel Foucault, the kind of author who delights in keeping readers guessing (Foucault, 1975): Berger introduces a concept, modernity, obviously of central importance in his work, tells us that it's frequently misused, hints at the bare beginning of a definition, and then we're on our own. Perhaps I am taking the concept modernity – sometimes exaggerated to a condition of hyper-modernity (Berger, 1986) – too seriously, but given the circumstances in which it occurs and the ways it's used, that seems extremely unlikely (Berger, Berger & Kellner, 1973).

Elsewhere Berger defines modernity as an "agglomeration of technical, economic, social, and cognitive elements, all of which are available to the historian and the social scientist" (Berger & Kellner, 1981: 147). This extraordinarily vague statement reads like an old-fashioned definition of culture taken from an anthropology textbook published in the 1950s. It includes just about everything, and it is substantively coarse-grained and scattered rather than internally patterned in an informative way. Moreover, everything is presented as equally important.

Faced with Berger's stubborn reticence regarding a suitably workable definition of modernity, we may decide to turn to thinkers and ideas that have had the strongest influence on Berger as a sociologist and hazard a tentative inference based on their ideas. First and foremost, as we have already made clear, Max Weber is the classical theorist whose work has had the greatest influence on Berger, especially in terms of the rationalization of social systems.

Weber identified four types of rationalization, but he was most keenly sensitive to what we have termed functional rationality, frequently presented in German as Zweckrationality. Functional rationality thoroughly pervades all modern and modernized societies and is best reckoned in terms of exchange value, determined by supply and demand (Horkheimer & Adorno, 1947; Henderson & Parsons, 1947).

Weber also gave an intellectually privileged place to value-oriented rationality, geared toward the specification of use value and commonly rendered in German as Wertrationality. The value of things in themselves, or use value, is much more difficult to measure in a plausible way than exchange value, which can be readily expressed in exchange theoretic terms such as dollars (Henderson & Parsons, 1947; Horkheimer, 1947/2013; Blau, 1964).

If, as Berger is inclined to do, we attend only to cause-and-effect connections, we miss much of the man-made world – much of culture – because we ignore the difficult questions raised by gauging Wertrationality, meaning that we typically lack a suitable metric for use value (Horkheimer, 1947/2013).

Producing things that can be measured in exchange theoretic terms is essential to serving the interests of capital accumulation and the exploitation of man by man. However, when an object or process of concern is rationally evaluated in terms of Wertrationality or use value, the interests served are potentially universal, and a basis for measuring them may defy specification.

Nevertheless, as Berger tacitly construes things, exchange value is all-important, and use value is something he does not effectively address. This is especially evident in his construction and discussion of fifty hypotheses concerning the nature of capitalism contained in *The Capitalist Revolution* (Berger, 1986). The underscored character of functional rationality places Berger in the company of American pragmatists such as John Dewey and William James (Horkheimer & Adorno, 1947/2007).

If we take the foregoing to mean that the one-dimensional emphasis that Berger places on functional rationality makes it central to his understanding of modernity, we may be making progress toward a definition. For Berger, the Weberian pervasiveness of rationality, especially functional rationality, seems to be the primary defining characteristic of modernity. Still a one-dimensional definition of modernity seems unhappily incomplete, especially since this means that Berger himself has only a one-dimensional understanding of the concept. Besides, we are still talking about Weber's understanding of modernity, not Berger's own (Berger and Luckmann, 1966; Berger, 1967).

While Berger has a strong affinity for Weber, he also acknowledges that Weber's work, like that of all other sociologists, is time dependent and, therefore, not well suited to addressing many important contemporary issues (Berger, 1986). So how seriously should we take rationality as the central component of modernity, much less as the fundamental substratum on which the concept rests within its overarching cultural context?

Berger insightfully notes that the ambiguously powerful nature of modern science-based technology can be shaped for uses that are, by one standard or another, good or evil. This assures us that one characteristic of modernity is the dynamic tension between liberation and re-enslavement (Berger, 1986). Sounds serious to me, and it is, but as far a definition of modernity is concerned, it's just another brilliantly conceived hint, this one based on conflicting outcomes of the use of science-based technology.

At this point it's worth repeating the judgment that in less formal discussions, the meaning of modernity might be taken as sufficiently self-evident that uncertainty and vagueness would be inconsequential, and lack of detail might be a welcome simplification. In a work of technical sociology, however, this is hardly a usefully informative way to proceed. That's why we continue with this tiresome search for a legitimate definition for modernity as Berger uses it.

Furthermore, Berger has argued that modernity is in a state of crisis (Berger & Kellner, 1981). He once again fails to develop the meaning of an interesting assertion, but if a phenomenon as central to his sociological interests as modernity is headed toward some sort of ontological or epistemological catastrophe, it would be extremely useful if Berger explained just what was headed for disaster of apparently substantial magnitude. As it is, modernity seems to be in some way coupled with what Berger terms "modern secular rationality." But this is really just a repetition of our previous reference to Weber, and, as such, adds little to what the reader has already surmised, especially since we don't know what a crisis or catastrophe in "modern secular rationality" or "modern technical rationality" would mean or would look like (Berger & Kellner, 1981: 7).

To make matters more difficult, Berger has concluded that the process of developing sociology, since Comte first put pen to paper, has been organized around issues concerning the nature of modernity (Berger, 1986). Perhaps so, but even if sociology and modernity are involved in some sort of mutually implicative developmental flux that will continue indefinitely, a simple, straightforward, reasonably adequate-for-the-time-being definition strikes me as an eminently reasonable request (cf. Wagner, 1994).

When Berger tells us that a "focus on modernity implies an effort to see contemporary society as a whole, in itself and in relation to other societies," he generates more confusion than clarity (Berger & Kellner, 1981: 8). After all, what does he mean by "contemporary society?" The United States? All social systems that have been thoroughly developed economically? "Contemporary society" could have an indefinitely broad range of meanings, and we have no reason to prefer one over all the rest. In at least a temporal sense, Haiti and Chad are comparably contemporary.

In *The Capitalist Revolution* (Berger, 1986) we are told that industrial capitalism and industrial socialism are modes of economic organization that Berger construes as "twin manifestations of modernity." However, the meaning of this puzzling judgment is not explained, but what else is new? Add to that the assertion that "While industrial capitalism is one version of modernity, it is not coextensive with modernity" (Berger, 1986: 71), and it seems reasonable to argue that we have gained nothing in clarity but we have added still more confusion. Moreover, it's a bit disconcerting at this stage, to learn that there is more than one version of modernity.

Given the widespread expectation of cross-generational upward mobility and the profusion of prosperity, Berger characterizes America as "a specifically virulent case of modernity as such" (Berger, 1986: 49). Why "… as such …?" This one eludes me in many ways. I have to admit that I don't know what he means.

An interesting observation that tells us something useful about modernity without actually defining it appears in *A Far Glory* (Berger, 1992: 89): "Modernity is a giant movement from fate to choice in the human condition." Insofar as modernity includes urbanization, technology-intensive economic development, mass communications, and a market economy, this certainly seems pertinent, even though incomplete as written. Nevertheless, the super-imposition of choice over fate suggests another way in which functional rationality would broadly manifest itself.

Since both capitalism and socialism represent an attempt to rationally organize the process of production toward a desired means-end state, it does seem reasonable to take "rationality" as a defining characteristic of modernity. Though the evidence in support of rationality as the foundation of modernity continues to accumulate, this is still a tentative inference, not something explicitly stated by Berger. Nevertheless, it's consistent with Berger's invocation of Weber's concept "rationalization" or "rationality" or "modern rationality." At this point it seems foolish to reject as inadequate the evidence that has accrued consistent with functional rationality as the central defining characteristic of modernity. This still seems incomplete, but it may be the best we're going to do, especially if we join functional rationality with choice.

When all is said and done, I think we can simply say that we're disappointed that Berger didn't do a much better job explaining a concept that has made a substantial contribution to defining the borders of his sociological career. After all, in his recent book, *The Many Altars of Modernity*, Berger (2014) tells us that the relationship between religion and modernity has been foremost in his thinking throughout his still productive life as a sociologist. Finally, moreover, Berger gives us an explicit definition of the concept that still seems elusive to him:

> Modernity can be most succinctly defined as the changes brought by science and technology – an ever-accelerating process, with consequences affecting ever more areas of human life. In a way, modernization means a sort of expanding urbanization, so that more and more people throughout the world come to breathe "city air," even if many of them still live in places that are not or not yet, cities. In a way the whole planet becomes an enormous city (Berger, 2014: 5).

Even this definition, as simple – even crudely simple-minded – as it is seems is obviously unsatisfactory, maybe even sloppy. The first sentence very briefly sketches a bare-bones definition for modernity. The second addresses the connection between modernization and urbanization, but modernity itself is yet again ignored. The third sentence is apocalyptic but not in a way that further informs us as to the nature of modernity. Furthermore, the definition seems unrelated to – in some cases even at odds with – the previous references to modernity as we have recorded them. Such an important concept, but treated so shabbily. "Slipshod," "sloppy," "indifferent" – very disappointing!

Mead got around difficulties in defining "universals" by addressing them functionally: phenomena that are different but to which we respond in the same way (Mead, 1934/1967). Perhaps this adroit use of language is too clever by half, but it provides us with an approach to modernity that is the most satisfying Berger offered: "modernity … undermines all the old certainties; uncertainty is a condition that many people find very hard to bear; therefore, any movement (not only a religious one) that promises to provide or to renew certainty has a ready market" (Berger, 2014). I find this functional approach especially useful in trying to define modernity. It is an assessment informed by the work of Mead, Tönnies, and Durkheim, and it's quite illuminating, even though it's not, strictly speaking, a definition.

The relationship between modernization and modernity

There are numerous places in *The Homeless Mind* (Berger, Berger & Kellner, 1973), where Berger seems to be carelessly using modernization and modernity as if they were interchangeable ideas. The previously quoted definition of modernity from *The Many Altars of Modernity* (Berger, 2014) might be construed as consistent with that interpretation, but it would be a stretch, and out of synch with so much else that Berger has written.

In the social sciences, moreover, modernity has most often been treated as a consequence of modernization, or a concept that stands in a reciprocal relationship with modernization, and we are treating it as such. Modernity, then, ill-defined though it remains, is a dependent variable with modernization as independent. Furthermore, the relationship between modernization and modernity introduces what econometricians call endogeneity issues, meaning that causal influence may flow with equal or variable strength in both directions (Paxton, Hipp & Marquart-Pyatt, 2011).

Modernization, too, has a value-laden sound, especially since it is often associated with the closely related concept of development, an idea that suggests developed and, too often pejoratively, undeveloped or less-developed societies. However, as used by Berger, modernization is definable in value-neutral terms, though it's quite clear that for him modernization is a difficult process that offers magnificently favorable outcomes in terms of the gains in material prosperity that he values so highly.

In principle, modernization is also intended to be measureable, though operationalizing the definition offered by Berger would require as much ingenuity and nerve as my tentative definition of modernity. Berger defines modernization as "the institutional concomitants of technologically induced economic growth" (Berger, Berger & Kellner, 1973: 9). There's not much there, but at this point, after our struggle with modernity, that is no reason to be surprised.

Whatever the institutional concomitants turn out to be, and which institutions are complicit in the process, we can readily see that we cannot take the concept modernization as just rendered and apply it to a social system and say that it is or is not modernized. According to Berger, however, once we better understand his interpretation of modernization, we will be able to use the definition in a way that will enable us to place different societies on a continuum of greater and lesser modernization (Berger, Berger & Kellner, 1973).

Whatever Berger's intentions, this unmistakably implies that modernization can be construed as an ordinal level variable, and perhaps interval or ratio. Berger's bare-bones definition, however, presents some painfully obvious difficulties with respect to measurement, and therefore is at odds with the creation of new variables (Lockhart, 2014). One might imagine that discrete organizational typologies, once the mainstay of research on complex organizations, would yield the not-particularly-informative variables that would follow from Berger's measurement efforts and his unspecified endeavors in statistical analysis (see, for example, Etzioni, 1964; Maguire, 2003; Scott, 2013). At this point there is too little information to make a well-informed judgment.

6 Socially constructing modernization, modernity, and our social world

Berger owes a significant conceptual debt, that he gratefully acknowledges, to George Herbert Mead. Mead's understanding of the essential role of shared language in creating the social world as a predictable and mutually interpretable set of relationships that reproduces itself from one generation to another is most conspicuously evident in his explanation of "the generalized other." While Berger's work on the creation of distinctive selves in varied social settings is fraught with originality, one wonders if his insights would have been nearly so keen without Mead's work in *Mind, Self, and Society.*

However we construe modernization and modernity, whether measured or treated as qualitative entities, it is important to recognize that insofar as they are manifest in changes in the way things are ordered and accomplished, they are products of human thought and activity (Berger & Luckmann, 1966). Changes in the social world, whatever form they may take and whatever definitions they may entail, are due to the way that human beings understand the nature of their circumstances and their expectations for consequences due to specific modifications (Berger, 1967).

Actions that enable ideas to take the form of concrete, perhaps measurable reality are inevitably due to purposefully directed human activity. Our ideas take form externally. To paraphrase Berger, we are dealing with the objectifications of subjective processes and meanings by which a socially structured world manifesting modernity or other forms of social organization will be created (Berger & Luckmann, 1966; Berger, 1967). In the abstract, these processes and meanings are part and parcel of what we have termed modernization.

At this point, we see that three of Berger's most basic explanatory concepts are simple enough: institution, first and foremost, followed by technology, and economic growth. Furthermore, throughout the rest of what he has written to date, Berger continues to use these very basic, very abstract ideas in much the same offhanded way, though with respect to modernization and modernity technologically-intensive production quickly takes pride of place as the motive force in Berger's set of concepts (Berger, Berger & Kellner, 1973; Berger, 1986).

Modernization, it is useful to make explicit, entails a re-organization of society. A legitimate question, then, one with which Durkheim was concerned throughout his life (Fournier, 2012), is how is a particular, more or less stable form of social organization possible. The best answer I have seen, one that follows from our earlier rendering of the dialectic, is that social organization, modernization included, is a human production, something that man creates in his continuing process of expressive externalization, objectivation, and internalization. More simply and specifically, as we have noted above, commonsense knowledge produced by man in everyday linguistic interaction makes it so in that it provides the social building blocks located in our intersubjective social stock of knowledge (Berger, 1963).

A particular form of social organization or social order cannot be understood as a manifestation of the nature of things, but it exists only as a product of expressive human activity (Berger & Luckmann, 1966: 147–167). Much as with man's self-production, production of a specific kind of social organization – a social order – is always, and by its very nature, an active and ongoing social process (Berger, 1967).

Taken-for-granted predictability

In this vein, the fundamental concept institution seems to have acquired a taken-for-granted character that tends to minimize its importance simply because it has become so familiar, whether the concept is used properly or not. Technology-intensive production, however, could not take place without the formation of social institutions that predictably direct the activities of those involved in productive work (Berger, 1963; Berger, Berger & Kellner, 1973).

As such, institutions are not merely collections of ideas that we would like to see in action. Instead, they have a forceful, thoroughly objective facticity that regulates performance, including technology-intensive production, in a fashion suitable to attaining the intended ends. The fundamental importance of the concept "institution" prompted us to discuss it at length early in this manuscript. We will encounter it in a variety of settings, and its sociological importance cannot be exaggerated (Berger & Luckmann, 1966; Berger, 1967).

At this juncture and throughout our presentation, it is useful to bear in mind that institutions, in all their objective facticity, would not be possible without a sign system that has the power and versatility of language. After all, in *The Social Construction of Reality* (Berger & Luckmann, 1966) Berger begins his discussion of institutionalization with an isolated pair of actors who may or may not be able to communicate in any way except through the construction of typifications that they cognitively attribute to each other. Institutionalization as we understand it becomes possible only with the introduction of a third actor, one who can communicate in the same way and who, through the usual means, participates in the formation of a succeeding generation.

It is only for this second generation that behaviors that were first exhibited and that became habituated for its predecessors acquire the character of being part of the shared social world. The second generation actors did not see the initial construction of these nascent institutional forms, so they take them for granted and do not question their provenance.

Berger, I'm sure, would be the first to admit that a process such as that just outlined is not something that actually happened as described, that it was ever concretely real. Its instructional purpose, however, seems self-explanatory, and it provides us with a time span over which unspoken typifications can plausibly begin to manifest themselves into a rudimentary language which will permit transmission of information to succeeding generations. It is only through language that the reality of everyday life takes shape, and it is only through language that we come to apprehend it (Berger & Luckmann, 1966).

Mead's (1934/1967) explanation of how linguistic gestures become significant symbols – mutually interpretable by the speaker and the spoken to – is invaluable in this process. It is only through ongoing social interaction that rudimentary linguistic symbols become understood as words, components of a shared language, and human beings acquire the ability to intelligibly converse. This immensely facilitates the processes of habituation, externalization, objectification, and internalization, the procedures that make possible the social production of institutions.

In effect, institutions are objectified language, and they persist as such, while we contribute to their efficacy and stability through our routinized behavior and objectified thought (Berger & Luckmann, 1966). In general, the longer institutions last as objectifications of our externalized and purposefully thought out meanings, the more durable they become. With time reckoned in terms of generations, this is not difficult to see. In this regard, the following quotation from Berger is especially pertinent:

> [L]anguage is capable of becoming the objective repository of vast accumulations of meaning and experience, which it can preserve in time and transmit to following generations (Berger & Luckmann, 1966: 37).

This brief quotation is pregnant with interesting interpretations pertinent to the creation and nature of institutions. First and foremost, it is obvious that a common stock of knowledge is possible only because of the persistently objectified power of language. Following from that, the meaning-laden linguistic material from which institutions are structured is located in the common stock of knowledge, possible only because language does, indeed, take a durably objectified form, meaning that it transcends specific situations.

Furthermore, institutions are possible only because the common stock of knowledge permits reciprocal typifications. Through the kind of social interaction made possible by language, two or more actors come to have durably complementary expectations of the kinds of behavior that specific actions will beget from others.

A simple example, the kind George Herbert Mead used very effectively (Mead, 1934/1967), might be a pitcher throwing his best fastball, the batter hitting a weak ground ball back to the pitcher, the pitcher throwing the ball to the first baseman, the batter is out, and the ball is thrown back to the pitcher who prepares to throw another pitch to another batter. Each player was able to take the role of the other, and as a result they knew how to behave, each in response to the others.

When we acknowledge the reciprocity of perspectives involved in playing baseball, it suggests the fact that for a human being to play an institutionalized role is always contingent on others recognizing him as occupying that role (Berger, 1967). After all, if the pitcher had fielded the weak grounder and thrown the ball to the left fielder, the institution of baseball, in that instance, would have been hopelessly and inexplicably compromised.

From this simple example we immediately see that as with any institution, baseball embodies roles played by actors who understand the rules that govern each role, theirs and those of the other players. That these roles are products of secondary rather than primary socialization is apparent in the ease with which one can step out of one role into another, or leave the game altogether. This in no way diminishes the institutionalized nature of the roles. After all, in the case of baseball, the roles have persisted for more than a century, but the degree to which the actors have been socialized into them is not so strong as to make them life-changing and permanent parts of one's identity (Berger, 1963). The case of established professional players may be an exception.

Anyone familiar with the game of baseball will immediately and without reflection understand the sequence of events just described. He will also be able anticipate a broad range of alternative outcomes following the pitcher throwing his best fast ball, perhaps a line drive into the seats just over a short right field fence. Baseball is a comparatively simple institution with fixed rules that combine in a surprisingly complex variety of ways making a particular kind of social order possible. As long as play is by the rules, behavior is legitimated, meaning that it is real and normatively correct (Mead, 1934/1967).

However, the assumption that knowledge of how baseball is played is invariably part of an all-inclusive common stock of knowledge is sometimes embarrassingly erroneous. Anyone who has taught a graduate class with males and females, including international students from a broad range of countries, and tried to use a baseball example – say the strength of the relationship between number of home runs and annual salary controlling for number of RBIs (runs batted in) – will certainly understand this. If one were sufficiently foolish to try to explain the game in a minute or two, the complexity of the comparatively simple institution of baseball as it has evolved for over one hundred years would quickly become apparent.

Participation in a common stock of knowledge is something that we take for granted, but our assumptions need to be well founded or we risk misunderstanding. Part of the common stock of knowledge, thus, is awareness of its content. For those who have grown up playing and watching baseball, the objectification of rules expressed in language into a complex but unambiguously ordered set of behaviors is easy to see. While baseball is sufficiently complex that it resists being taught in a brief session or two, the objectification of its rules in language and behavioral repertoires assures that it will persist, with one generation after another being socialized into its established ways of behaving (Berger & Luckmann, 1966).

Each of us, moreover, is familiar with a very large number of institutions of greater and lesser importance and complexity. Our memories and learned repertoires of physical performances are repositories for an enormous collection of institutional templates.

Perhaps it would be useful if baseball and other leisure-time objectifications had more participants and more inveterate fans. This might compensate for what Berger, in *The Homeless Mind* (Berger, Berger & Kellner, 1973), unhappily characterizes as the under-institutionalized nature of the private sphere of life.

Given changes in our economic order that have unexpectedly increased the need for long hours of work just to get by, this is no longer the problem it seemed certain to become in the 1960s and early 1970s (Schwarz, 1998). Nevertheless, boredom and purposeless inactivity have certainly not entirely disappeared from the private sphere.

Objectification or objectivation

As in the foregoing, any discussion of the formation of institutions is likely to make frequent use of the concept of objectification. As far as I can tell, in *The Social Construction of Reality* (Berger & Luckmann, 1966) Berger uses the term objectification interchangeably with Weber's concept of objectivation, whereby meanings are made substantively real in the sense that they are capable of transcending a specific context. In fact, Berger uses objectivation before he defines it (again no surprise), and in *The Sacred Canopy* he uses it even earlier (Berger, 1967: 9), and leaves it undefined throughout.

Why Berger sometimes uses objectification and other times uses objectivation, when either used consistently would do is not explained. Perhaps he wanted to make emphatically clear that Weber understood the use of thoughts, meanings, ideas, and language in ways that lent themselves to creation of objectively real social structures. When Berger was convinced that he had made the point that Weber understood the place and importance of this process and of questions as to how it could be accomplished, he no longer felt the need to use the German word objectivation instead of the English equivalent objectification. Perhaps, but this is certainly weak speculation on my part.

It has also occurred to me that Berger uses objectification when referring to behavior expressed as language, but when expression is in terms of something more materially substantial he uses objectivation. Again, however, I'm just speculating. Besides, Berger holds that objectivation refers to both physical and mental products of externalization. I take mental products to refer to less materially substantial products, particularly language, and this speculative exercise comes to an end (Berger, 1967).

End notes in *The Social Construction of Reality* (Berger & Luckmann, 1966) and *The Sacred Canopy* (Berger, 1967) contain very brief accounts of the Hegelian and Hegelian/Marxist origins of objectivated and objectified. However, I did not find the end notes helpful and I doubt that more than a few readers would. Adding to the mystification is Berger's use of both objectivated and objectified in the same paragraph and even the same sentence with very clearly the same meaning (Berger & Luckmann, 1966: 68).

Whatever Berger's reasons for using objectification and objectivation interchangeably in *The Social Construction of Reality* (Berger & Luckmann, 1966), in *The Sacred Canopy* (Berger, 1967) he restricts himself to objectivation. No reason for this is given, but at the very least one may find that this simplification diminishes terminological confusion. Again, this is just weak speculation on my part.

Subjective meanings as objective facticities

In any case, we may use this distinction to bring us back to an all-important question: "How is it possible that subjective meanings become objective facticities" (Berger & Luckmann, 1966: 18)? We have given substantial attention to answering this question, but it remains difficult. Among other reasons, as Berger sees it, the difficulty reflects the fact that it is very easy to forget the dialectical character of the relationship between what we do and how we think (Berger & Luckmann, 1966).

This holds whether we use objectification or objectivation. Besides, when we first encounter assertions such as "An institutional world, then, is experienced as an objective reality" (Berger & Luckmann, 1966: 60), they seem, at best, counterintuitive, and maybe even nonsensical. This is genuinely difficult material.

Beyond that, whichever word we prefer, objectivation or objectification, Berger's admonition with regard to avoiding reification is well taken. Given the forceful facticity and power of determination possessed by social institutions it may be difficult to remember that they are human creations, and not regard them as if they were aspects of the world of nature (Berger, 1963). Nevertheless, were it not for reification, there would be neither institutions nor roles that constitute them (Berger & Luckmann, 1966). An interesting set of circumstances, indeed.

By all means, remember reifications for what they are, but don't fail to understand how institutions are in fact products of meaning-laden social interaction among human beings. After years of dealing with sociological issues and related material, I still have to stop every now and then and think it through.

Institutions and social roles

I think that Mead's (Mead, 1934/1967) account of interlocutors simultaneously taking the role of the other captures the dialectical character of effectively participating in the social world in a way that is clearer and more compelling than Berger's initial rendering in *The Social Construction of Reality* (Berger & Luckmann, 1966). It is clear from Mead, moreover, that the possibility of a role-free existence simply does not exist.

The force and clarity of Mead's presentation holds even though it is offered in a casual, matter-of-fact, spontaneous sort of way (Mead, 1934/1967). I am not denying the existence of a dialectical relationship much as Berger presents it in *The Social Construction of Reality* (Berger & Luckmann, 1966), but I am saying that Berger's account is, at least at first, not particularly well executed. It may very well be that it's just the offhanded spontaneity that characterizes Mead's presentation that makes it more compelling.

However, I have subsequently found it mnemonically useful to follow Berger's lead and think of man as producer and the social world as his product in terms of Berger's three dialectical moments: externalization, objectification, and internalization. Corresponding to these three dialectical moments, Berger gives us the following: "Society is a human Product. Society is an objective reality. Man is a social product" (Berger & Luckmann, 1966). I find this helpful not only because I have memorized it, but because the more often I read it the clearer it becomes.

Though Berger rightly emphasizes that the three moments occur simultaneously rather than sequentially, his version of the dialectic, while substantively brilliant, still seems stilted and lacking in conceptual dynamism. I don't doubt that he is fundamentally correct, just that his explanation is not stylistically forceful. Externalizing one's being in the social world and simultaneously objectifying it as bare facticity that is then internalized is not a hopelessly complex formulation, but for me it lacks the energy and thrust that one expects of a dialectical relationship (cf. Marx & Engels, 1846/1998). Maybe it's just a matter of taste.

However, Berger's discussion of the dialectic in the first pages of *The Sacred Canopy* (Berger, 1967), though substantively the same, is stylistically far better. If I had read *The Sacred Canopy* (Berger, 1967) before *The Social Construction of Reality* (Berger & Luckmann, 1966), and had not been unduly eager to put pen to very preliminary paper, I might have avoided my less flattering comments directed at the explanation of the dialectic as it appears in *The Social Construction of Reality* (Berger & Luckmann, 1966). Nevertheless, my comments are accurately honest as written, though only a year separates publication of the two books, and both are theoretical in character.

In any case, in *The Social Construction of Reality* (Berger & Luckmann, 1966) I think that Berger does a much better job when he discusses the role of significant others in a child's primary socialization. Here, too, he uses the same dialectical formulation, but much more effectively.

The child soon understands that he has been born into an objective social world and the influence of significant others enables him to find his place therein. It is important to emphasize, moreover, that this is a linguistically rich process. Externalization, objectification, and internalization are all accomplished in terms of language.

Berger's discussion is especially insightful when he explains that a child born into an identifiable social class in a society's stratification system has a class-based understanding of the world, but that this is subject to the interpretation given it by his significant others, especially his parents. A child born into a particular class not only objectifies and internalizes the point of view intrinsic to that class, but he internalizes it in terms of the distinctively peculiar slant given it by significant others with whom he identifies. This slant, moreover, is another product of language, as is the child's class-based understanding of the world.

Berger's judgment that the reality of cross-class differences in the way we understand the world should be interpreted with reference to within-class differences is an acute observation that merits close scrutiny. Sociologists with a quantitative bent will immediately see the applicability of the logic of multilevel analysis in situations such as this (Bickel, 2007; Hox & Roberts, 2010).

Given the foregoing, as the child internalizes significant others' understanding of him and he makes it his own, he begins to internalize their subjective understanding of the world. In this way, the child takes on the significant others' roles and the affect with which they invest their circumstances, and he does so using the language he comes to share with them. This process is much like learning the rule-governed activity of all the positions on a baseball team, but it is a good deal more complex.

The child identifies with significant others and he identifies himself accordingly. In a brief but brilliantly insightful passage, Berger takes the position that identity formation inevitably includes the following:

> It is possible to sum up the dialectic formation of identity by saying that the individual becomes that which he is addressed as by others (Berger, 1967: 17).

This is a linguistic process through and through. The child becomes one with his interlocutors, and his identity gains coherence as he internalizes the roles and attitudes of the group, and comes to appreciate what its members have in common, including motivations that manifest themselves subconsciously. The enhanced linguistic prowess required by this ongoing exchange is something the child acquires coterminously. In the process, the child acquires a taken-for-granted world, one that exists without question as real and right and that is consistent with his subconscious motivations.

The generalized other

The motivations and constraints that members share take the form of what Mead terms the generalized other (Mead, 1934/1967), a set of guidelines that come to regulate the child's behavior. In an abstract sense, when the child internalizes the generalized other he internalizes the world, or at least the part of the world that he and his significant others call their own in a taken-for-granted way. Quoting Berger with regard to the generalized other the following is useful:

> Society, identity, and reality are subjectively crystallized in the same process of internalization. This crystallization is concurrent with the internalization of language. Indeed ... language constitutes both the most important content and the most important instrument of socialization (Berger & Luckmann, 1966: 133).

With the internalization of the generalized other, the child has become a member of society. Perhaps it's a modernized society, perhaps not. In abstract theoretical terms it makes no difference. The generalized other, however, is not a once-and-for-all entity. It changes with the rewards, penalties, and demands of the world in which it is located, still serving the same regulatory purpose and still being an ongoing social creation, providing the subconscious motivations inherent in participants' location (Berger & Luckmann, 1966). Fundamentally, its source of directive material is the common stock of knowledge (Berger & Luckmann, 1966).

While I certainly agree with Berger that the generalized other is what he terms a nomic structure, meaning that there is a shared nomos or active repository of commonly held, interpretably ordered meanings, I don't share the "everything is all right" interpretation that Berger offers as coterminous with the nomic structure (Berger & Luckmann, 1966).

Children who are born and raised in homes that are, with barely a moment's respite, wildly chaotic know, in some sense at every age that everything is *not* all right. The older they get, the more evident and unrelenting is their awareness of chaos and trouble. Nevertheless, I can think of no reason why the generalized other cannot take shape in a destructively unsatisfactory world. It will, of course, manifest the character of the world in which it is formed.

The generalized other that takes shape under these unhappy circumstances may be much more uncertain and home-bound than most. Experiences outside the home may or may not be quite different, but whatever the outside conditions, most of the child's primary socialization, including his acquisition of subconscious motivations in a taken-for-granted world still takes place within the home amid chaos and the perception that things are not all right.

It may be the case that this results in a generally unsatisfactory life, through adulthood until death, but the "everything is all right" admonition seems unrealistically hopeful and misleading. Efficacious primary socialization does not presuppose a wholesome or even more or less stable environment. Maybe you had to be there. My experience tells me that even a chaotic environment, in its own unfavorable and perhaps difficult to discern fashion, is organized predictably and in terms of a distinctively turbulent order. The emergence of mutually identifiable roles played by participants still represents a process of institutionalization. Though the roles acquired may be in various ways debilitating, they are still social roles, the kind that we can't simply step in and out of as we please.

Society as man-made and inherently precarious

We have made the observation that institutions in all their force and complexity are taken for granted and that references to them are commonplace but often betoken failure to appreciate their nature and importance. I think we have now remedied that all-too-facile set of circumstances, and we have arrived at an understanding of institutions and the process of institutionalization that does justice to the material.

We have also discussed social roles, how they are created, and the ways in which they are situated in the formation of institutions. We have given particular attention to the generalized other, a concept due to Mead that is the repository of that part of the taken-for-granted social world to which one belongs (Mead, 1934/1967).

While I sometimes took issue with Berger's account, his contribution is decisive. His understanding of the emergence of institutions is thoroughly developed, and rarely misses the mark. My objections pertain to details. Berger's indebtedness to Mead is substantial, but Berger's original insights are undeniable. This, I think, is a point usefully made in the following from *The Sacred Canopy*:

> Society is a dialectical phenomenon in that it is a human product and nothing but a human product, that yet continually acts back upon its producer. Society is a product of man. It has no other being except that which is bestowed upon it by human activity and consciousness. There can be no social reality apart from man (Berger, 1967: 3).

I can't imagine a clearer, more persuasive statement than that, but I can imagine struggling to understand it the absence of a clear understanding of the dialectic as such. In truth, though we may not see it at first, the process being described is reification. As we know, Berger cautions us against forgetting that the institutionally embodied social world is the product of reification rather than being born of the natural world. However, we need also remember that without reification there would be no social world (Berger, 1963 & 1967).

Perhaps it's too obvious to merit additional discussion, but I think it's useful to again make perfectly clear that the foregoing accounts of the linguistically heavy-laden dialectic and the creation of institutions and social roles are necessary only because man, in contrast to other creatures, is born in a radically unfinished fashion. As a result, human beings must create the world or worlds they occupy. They do so using the procedures and practices we have just been discussing, and they maintain or change that social assemblage in just the same taken-for-granted way.

To put this a bit differently, man differs sharply from other creatures in that he is born instinct-poor and with a bare modicum of biologically inscribed practices (Berger, 1963). He is singularly lacking in concrete reality-building procedures, thus making the externalization-objectivation-internalization dialectic essential to the making of social worlds (Berger, 1967).

Other animals may subsist for indefinitely long periods without changing biologically or socially. Man, however, with his openness, facility with the dialectic, and mastery of language may change rapidly and in wholesale fashion so that from century to century (if not faster!), his social and cultural worlds are dramatically different, sometimes almost unrecognizable as they follow one after the other (Berger & Huntington, 2002).

Clearly, in contrast to other animals, replete with regulatory instincts and pre-programmed with goal-directed behavior patterns, man and his social products constitute a precarious set of circumstances. The world of human beings is made and remade without a pre-given template, and with no guarantee that, from generation to generation, errors will not occur (Berger, 1967).

7 Production through science-based technology

In Berger's scheme of things, as presented in *The Homeless Mind* (Berger, Berger & Kellner, 1973), the primary motive force underlying modernization is science-based technology applied in a way that generates increases in conventional economic measures such as gross national product per capita, gross domestic product per capita, and per capita income (Berger, 1986). As such, Berger views modernization as presupposing a reasonably stable social setting, with the moderating influences that come with a predictable political environment made manifest through functionally rational, massively bureaucratic regulation of essential activities (Berger, Berger & Kellner, 1973).

Science-based technology and bureaucracy

It may seem odd to tie something as innovative and dynamic as science-based technological development to political bureaucracy, a staid and reputedly inflexible organizational form. Furthermore, in *The Capitalist Revolution* (Berger, 1986), Berger treats bureaucratization as a source of economic inefficiency, an inevitable and devitalizing characteristic of a socialist economy (Berger, 1986).

A great deal may have happened to change Berger's thinking in the thirteen years between the two books, and this span of time occasioned Berger's transformative discovery of capitalism. I think that Berger would argue, however, that the rationality inherent in bureaucratic organization is essential to the maintenance and functioning of a society in which technology can freely develop its full potential and then be applied in the most productive ways. Since Berger does not address the issue directly, we are left to speculate, an all too common set of circumstances with important issues in Berger's work.

Still, Berger's judgment that what Weber called legal-rational bureaucracies (Gerth & Mills, 1958), and Berger refers to as political bureaucracies or functional bureaucracies (Berger, Berger & Kellner, 1973), are massively essential in modernized societies is not without effective, even brilliant critics who caution restraint. Given my interests and background, Raymond Callahan's fascinating analysis *in Education and the Cult of Efficiency* is especially interesting (Callahan, 1960).

In addition, Peter Blau's insightful ethnography titled *The Dynamics of Bureaucracy* (Blau, 1955) was a very effective response to Ludwig von Mises's *Bureaucracy* (1944/1996), in which von Mises argued that public-sector bureaucracies, by their very nature, are incapable of effective responses to economic opportunities and are averse to change. Blau, in an ingenious application of exchange theory, demonstrated that effective innovation in response to changing circumstances is a common occurrence in public sector bureaucratic organizations.

Furthermore, one need not be a Marxist to appreciate Trotsky's indictment of soviet-style bureaucracy in *The Revolution Betrayed* (1936/2004). In Trotsky's view, Stalin's bureaucratization destroyed the Russian Revolution of 1917.

I mention the fact that the growth of modern bureaucracies has both defenders and critics only because Berger, ever the Weberian, takes it for granted that unbridled bureaucratization is inevitable in a modernized society, but he fails to acknowledge that this view is not universally shared without cautionary arguments. I think that there are instances throughout Berger's work in which he uncritically adopts the truth of Weber's perspective, or an oversimplified caricature of Weber's perspective, even though it may include assumptions that are not warranted.

Berger's commitment to Weber

Nevertheless, Berger is one Weberian who, to his credit, has not held Weber harmless from all criticism. He may give Weber a pass when one is not warranted, but he's also capable of calling attention to Weber's errors and questioning the applicability of his work to present day circumstances, though not with the alacrity he does with Marx. In any case, Berger is no fan of the purposeless proliferation of bureaucracy, and neither was Weber (Roth & Wittich, 1978). Weber referred to the all-encompassing rationality ostensibly inherent in bureaucracy as "the iron cage," hardly a term of sociological endearment (Ritzer, 2004).

One might pause at this point and ask why Berger's adherence to Weber and his work is so strong. One reason, I would surmise, is that American sociology has long uncritically deified Weber, undervaluing Durkheim, and rediscovering Marx only in the last few decades. I don't know if Berger's interest in modernity was prompted by Weber's interest in the same issue, or if Berger found Weber while in pursuit of a theoretical framework for his own thinking about the social world. Sorting out which came first seems sure to be a thankless task, but in the absence of common concerns it seems unlikely that Berger would have developed a life-long interest in Weber's sociology. After all, they also shared an abiding interest in the sociology of religion (Weber, 1920/2010).

It's also the case, however, that in addition to being monumentally prolific, Weber sometimes, though certainly not typically, had a disarmingly uncluttered way of presenting what sociology was all about. For the newcomer to sociology, Weber's simple observation that social interaction is merely a process whereby people orient their behavior toward one another can be surprisingly illuminating, something I experienced decades ago (Henderson & Parsons, 1947).

Taking that simple observation farther, we may get the notion that "the web of meanings, expectations, and conduct resulting from this mutual orientation is the stuff of sociological analysis" (Berger, 1963: 27). For many of us this is fascinating, especially when we are new to sociology.

At that stage we may be forgiven for failing to notice the micro-level emphasis and nominalistic implications of Weber's position, rendering him still extremely interesting but theoretically limited. The excitement comes from the unfamiliar and really quite simple interpretative nature of some of Weber's observations. However, the degree to which Weber's sociology was actually interpretative is commonly exaggerated (Bickel, 2013). A great deal of what Weber wrote was focused on the macro-level. This does not preclude interpretation, but it is not the social place for studying face-to-face interaction or the meanings shared by individuals about micro-level issues.

For one who, like Berger, immersed himself in Weber's work in preparing his master's thesis and doctoral dissertation, Weber's voluminous sociology may become a lifetime preoccupation (Berger, 2011). The analyses of Marx and Durkheim may seem socially glacial in their macro-level compass and pace of development in comparison, devoid of the interpretative activity of concretely real people. Again, however, it is useful to recognize, that a great deal of Weber's sociology was focused on the macro level and contained less micro-level interpretation of the sort that Weberians, including Berger, so often celebrate (see, for example, Roth & Wittich, 1978).

The convergence hypothesis

Before the fall of the Iron Curtain, given that modernization or something much like it occurred in the Soviet Union as well as in the capitalist West, we can see why otherwise astute observers were intellectually seduced by the so-called convergence hypothesis. Proponents of this view, particularly the actors we earlier referred to as modernization theorists, held that similar developments in science-based technology and the institutionalized structures needed to support them would eventually make the capitalist West and the communist East more and more alike, meaning more and more modernized.

Furthermore, Jan Tinbergen (1956), who first used the idea of a convergence hypothesis, was especially interested in modernizing Third World countries. The fact that convergence has not occurred may come as no surprise, but we can see embodied in the idea of modernization the reasoning that gave rise to this expectation.

If we accept, with minor modification, the definition of modernization offered by Berger, the institutional concomitants of technology-intensive, bureaucratically moderated growth are of first importance. As a sociologist of knowledge, Berger is especially interested in modernization as it leads to a particular cognitive content and manner of reasoning with identifiable components (Berger, Berger & Kellner, 1973). At its most fully developed, this is probably best captured by the term "modernity," much as we have defined it in the absence of a useful definition from Berger. Berger's long-delayed definition is singularly unconvincing, offering little or nothing of value (Berger, 2014).

Modernity and the malleability of human beings

In this context, it is useful to again acknowledge that the process of modernization, culminating, we have assumed, in modernity, would not be feasible if human beings were not fundamentally malleable, as we have repeatedly emphasized. The social and cultural transformations associated with wholesale conversion from traditional to modernizing and modernized societies are sufficiently thoroughgoing that they would outstrip the adjustive capabilities of all but the most flexible organisms.

This point has been made more or less effectively by many sociologists and other authors. I think, however, that Berger's rendering in *The Sacred Canopy* (Berger, 1967), is the most elegant and compelling version that I have read. An engaging prose style is a valuable tool in making complex and difficult arguments, and in this regard Berger certainly has his moments, and other events, as well.

Legitimation: the first generation

Human beings, through the process of modernization, have undertaken to transform their world, building an alternative that is strikingly different from its predecessors. In the process, sometimes with substantial opposition, they have chosen a very different way of life, one that would be unavailable if they were less open to remaking their world and remaking themselves (Marx & Engels, 1846/1988; Berger & Luckmann, 1966; Berger, 1967).

It is inescapably the case that much of the common stock of knowledge that saw people through everyday life in a traditional society will no longer be applicable. Remarkably, a new and different common stock of knowledge, a product of more recent social interaction in the form of externalization, objectification, and internalization, will come to prevail and provide behavioral guidelines, commonsense problem-solving formulas, and widely shared subconscious motivations suitable to a taken-for-granted modernized world. The new common stock of knowledge, too, will be construed as unproblematic and effortlessly shared with one's consociates (Berger & Luckmann, 1966: 87).

For those going through a process of active modernization such as that discussed in *The Homeless Mind* (Berger, Berger & Kellner, 1973), the changes in applicable knowledge and its value may seem to be an almost day-to-day concern. To say that this will be difficult is to understate the case. Little wonder that modernization has, in some instances, given rise to strong opposition (Berger, Berger & Kellner, 1973; Charlton, 2003). In the absence of widespread legitimating formulas that underwrite and render plausible the promise of processes of modernization, it seems certain that they would encounter a great deal more opposition than actually has been the case (Berger & Luckmann, 1966).

There is little doubt that some people will be left behind, unwilling or unable to make the new stock of knowledge their own. We see something like that today when older people refuse to use cell phones, are unwilling to use email, and complain that too many new and unneeded electronic gadgets are being introduced that serve no useful purpose. In truth, they often have no need for what they see as pointless innovations, and they lag behind in a way that may or may not be disadvantageous. Either way they are likely to be characterized as eccentric.

In *The Social Construction of Reality* (Berger & Luckmann, 1966) Berger includes a fairly lengthy, certainly original, and rather difficult discussion of the nature of legitimation (Berger & Luckmann, 1966: 92–128). I think, however, that Berger may generate needless uncertainty and a bit of confusion when he argues that for the first generation, those who created the objectified typifications that came to constitute institutions, no legitimation is necessary, but for subsequent generations legitimation becomes an issue. His point is that the first generation saw the institutions in a nascent state in which they made them and therefore know what they are for and why they work. Generations that follow, however, lack this direct producers' knowledge, meaning that they may have doubts.

Legitimation: across generations

That account makes sense, but it fails to take into consideration the powerful socializing processes that inculcate later generations into an existing social system. Throughout *The Social Construction of Reality* (Berger & Luckmann, 1966) and elsewhere (Berger, 1963, 1967, & 1986), Berger argues that participants take established institutions as being stamped with reality so firmly and persuasively that we commonly relate to them in a taken-for-granted manner as if they were part of the natural world. Legitimation, thus, "implies the establishment of symmetry between objective and subjective definitions of reality" (Berger, 1967: 32). The inclination toward this kind of relationship is so strong that Berger, as we have noted, admonishes his readers to avoid reifying social institutions, forgetting that they are man-made rather than naturally occurring.

It's true that socialization is never perfect, and in some instances we find wholesale failure of socialization. Nevertheless, Berger's argument on behalf of the objective facticity of institutions and their socializing force is so compelling that his comparison of the first generation with subsequent age groups makes its point weakly at best. Little wonder that thoroughly institutionalized activities are not more in need of remedial punitive and other corrective measures to maintain their integrity.

Bear in mind that the first generation knows the thoroughly mundane nature of what it has produced, while later generations may are subconsciously motivated to grant spiritually efficacious status to one or more institutional forms. After all, the institutions are right there, they're quite compelling, and those who know the pedestrian way in which they came to be are long dead. Who's to say that their legitimation shouldn't take the form of something supernatural, even god-like? The domain of the taken-for-granted is especially plausible, compelling, and convenient (Garfinkel, 1967).

It may be that an isolated society exists that has given rise to an overarching symbolic universe "and provides a comprehensive integration of *all* discrete institutional processes – that provides a cover so complete and effective that nothing exists that has not been legitimated" (Berger & Luckmann, 1966). This may seem farfetched and it is. But by positing the existence of a monotheistic God by whom everything that happens is intended and understood as part of a grand and good scheme of things, we may take comfort in the knowledge that everything, however horrible or wrong it may seem, serves a divinely benevolent purpose. In other words, everything that happens is legitimated (Berger, 1967). This is the kind of observation that prompted Berger's interest in theodicies, as previously discussed.

Legitimation challenged

Even in this seemingly all-encompassing case, however, it may very well be that when processes of modernization are introduced by faithless, aggressive men into countries where they represent alien forces and profane values, special problems of legitimation come to the fore. In instances such as this, cross-generational socialization may not be sufficiently powerful. The overarching symbolic universe itself, the most comprehensive level of legitimation that encompasses all institutionalized activity, may be threatened in its legitimating force by previously unimagined incursions that bear no resemblance to things as they are or to that which is known (Berger & Luckmann, 1966). The following from Berger captures this set of circumstances quite effectively:

> It is important, however, to understand that the institutional order … is continually threatened by the presence of realities that are meaningless in *its* terms. The legitimation of the institutional order is also faced with the ongoing necessity of keeping chaos at bay. *All* social reality is precarious. *All* societies are construction in the face of chaos. The constant possibility of anomic terror is actualized whenever the legitimations that obscure the precariousness are threatened or collapse (Berger & Luckmann, 1966).

Construction in the face of chaos! Anomic terror! Powerful statements that persuasively make the case that any system of legitimation is subject to destruction with disastrous social consequences. Insofar as modernization embodies processes that may possess this degree of social power, it will be terrifying to indigenous people living in what has been their taken-for-granted world.

Anomic terror may seem an unduly extreme means of characterizing the consequences of damage or destruction of a symbolic universe. However, the same process could be accurately characterized as anomie so strong and complete that one loses not only the satisfaction of routine participation in an established nomos, but he loses his understanding of what is real and even who he is. In effect, the nomos that rendered his experience understandably ordered and fraught with taken-for-granted meanings is rent asunder and ceases to transform chaos into coherence (Berger, 1967). Terrifying, indeed!

Legitimating modernization for those occupying recently modernized places may be extremely difficult. After all, how does one make things real and normatively correct – how does one legitimate! – the finality of death for people who have been socialized into a culture that includes the promise of eternal life? How does one legitimate frightening uncertainty for people who have thoroughly internalized the belief that they are never alone in a fundamentally benevolent universe? How does one legitimate individuation for people who have long been convinced that they are inextricably bonded together for all times and all places?

Berger neatly complements legitimation with the little-used word and process nihilation (Sartre, 1943/1993). At first look I suspected that this was another of Berger's neologisms, though in contrast to most I could not discern its meaning from its context, though it occurred to me that it might have something to do with nihilism.

As it turns out, "nihilation" is an established term in that it is included in some online dictionaries (see, for example, www.WordSense.eu Dictionary, retrieved 2016 from www.wordsense.eu). Oddly, however, from one dictionary to another, the definition is *exactly* the same, word for word. I can only conclude that this has something to do with the character of online reference material and the interests and limitations of those who maintain it.

Berger effectively defines nihilation in *The Social Construction of Reality* (Berger & Luckmann, 1966) to mean to exclusion or elimination of that which is not compatible with the prevailing symbolic universe. Nihilation casts out the normatively dubious and that which is not part of what is demonstrably real to ordinary participants.

Following Berger we may construe the process of modernization, particularly in Third World countries, as inevitably entailing an encounter – often a better word may be "clash" – between very different civilizations with very different symbolic universes. Imagine that for the first time, those living in a Third World country encounter a way of interpreting the world that differs from the interpretations they have taken for granted for longer than anyone can remember. They see, for the first time, that there are very different ways of completing essential tasks and thinking about ordinary circumstances and extraordinary ones, as well.

If those threatened with modernization can be shown or otherwise persuaded that the alternatives being offered are of demonstrably superior efficacy, they may incur doubts about wholesale rejection of modernization. The psychological, cultural, and physical processes will not be easy to accept and incorporate, but the promise of vastly improved living conditions may be sufficiently compelling that resistance will be sharply diminished. Again, socialization is never perfect.

Furthermore, the established symbolic universe may be maintained by skilled theoreticians who understand that long-term resistance is not likely to be effective. Thus proponents of modernization may acquire unlikely allies (Berger & Huntington, 2002). Those whose theoretical attainments have enabled them to achieve the status of keepers of the symbolic universe may very well find ways to include modernization in the province of legitimation that they creatively oversee.

Agents of legitimation: Christian and otherwise

The notion of theoretically accomplished keepers of the symbolic universe has an intuitively plausible appeal. In our modern world, priests, high-ranking politicians, upper-brow journalists, the obscenely wealthy, and those of exemplary intellectual, scholarly, and moral attainment quickly come to mind. Berger however, makes his presentation needlessly abstract when he does not provide concrete examples of the roles of these formidable actors as they vary from one social setting to another.

The same applies to the skilled theoreticians who may assist the forces of modernization. An articulate and culturally sensitive civil engineer may be worth more than a small army of Christian missionaries who have nothing to offer but their own peculiar brand of mysticism (Berger, 1986). The artfully productive engineer who can also talk a good game and demonstrate material improvements may offer an alternative symbolic universe that provides convenient opportunities for melding with the symbolic universe of the Third World country. Experts from the two sides need not be at odds, but may find more or less comfortable ways to connect, making the experts' jobs easier for all.

"Words of the cross" at odds with commonest of common sense

The missionaries, on the other hand, assuming they are Christians, offer the stark vision of a savior who, though he was the Son of God, was reduced to abstemious poverty and unceremoniously and brutally executed as if he were the basest of common criminals. This vision of the Lord is well known to Christians and most others in modernized countries, whether or not they worship Christ.

People in non-Western, less-developed societies, however, may respond much like those to whom the Apostle Paul preached the Gospel. They may find it simply inconsistent with taken-for-granted pre-suppositions – sharply counter to common sense! – to envision Jesus as both the Son of God and an impoverished martyr who was powerless when faced with death by crucifixion (Berger, 1992).

The self-sacrifice of a God or his offspring may be so inconsistent with a Third World country's common stock of knowledge that it seriously aggravates problems of legitimation. Difficulty legitimating the central tenets of Christianity was prevalent when Paul was preaching two thousand years ago, and the same may very well be true today in places where widespread understanding of the meaning of the "word of the cross" does not prevail (Berger, 1967). Recall, moreover, Berger's judgment that the institutions that constitute a society need not be logically consistent (Berger & Kellner, 1981). Though Berger does not explicitly say so, however, it seems virtually certain that participants in any social setting will wittingly and unwittingly strive toward a perception of consistency (Berger, 1967).

This difficult set of circumstances is explained with skill, insight, and remarkably acute understanding by Berger in *A Far Glory* (Berger, 1992). It forcefully but sympathetically brings to mind the fact that Berger is a sociologist of religion of the first rank. As such, he speaks authoritatively of the faith to which early Christians bore witness, and the sometimes contradictory taken-for-granted aspects of their culture.

Berger also makes clear that since the beginning, people who, perhaps disingenuously, called themselves Christians have sought ways to make their putative faith more consistent with the rest of their cultural ambience (Berger, 1992). This, no doubt, is one reason why there are so many different Protestant denominations, and one reason among many why modernized societies give rise to pluralism. When important components of a prevailing culture, components such as religion, are sharply at odds with others, this disjuncture may generate a good deal of psychological discomfort and foster efforts to diminish divisions and enable Christianity to comport with ostensibly changing times as seen from pluralistic doctrinal venues.

This task is especially difficult in a modernized society because its cultural ambience is rarely stable. From one generation to the next, culture may demand different form and substance from those who would keep Christianity in synch with the times or "up to date." How does one maintain a secure place for the sacred given difficulties such as these. Berger explains the demands intrinsic to this set of circumstances very well:

> In a society marked by pluralism the controls over interaction and communication with the outside have to be rigorous indeed. The slightest relaxation of these controls can breach the protective dam against the pluralistic infection, alternative definitions of reality will then flood in, and the precariously maintained taken-for-grantedness can collapse overnight. Therefore, the denizens of the subculture must limit contacts with outsiders to a minimum, avoid all unnecessary conversation, and equally

> avoid all media of communication originating outside. In other words, what must be established and maintained is a minitotalitarianism (Berger, 2014: 43).

Berger admonishes us that the wisdom and common sense of the age of the Apostle Paul are inevitably quite different from our own (Berger, 1992). As a sociologist of knowledge, this is something that is especially interesting to Berger (Berger & Kellner, 1981). Most of us, should we have reason to think of centuries past, may unself-consciously and erroneously take the common sense of then and now to be much the same.

Given the complexity of the tasks undertaken by experts from sharply differing symbolic universes, the task of melding parts of conflicting models will not be easy. When all is said and done, the sought-after and hoped-for theoretical compromises may not be forthcoming. At that point, belief in one symbolic universe or another may be a phenomenon whose coercive promise is located in the barrel of a gun (Berger, 2011).

One may understandably argue that force begets nothing but disingenuous lip service or fearful cowering rather than allegiance. However, if the institutions assigned with forcibly purveying belief are properly constructed and construed as agencies of socialization, their teachings may acquire the status of unquestioned truth (Berger, Berger & Kellner, 1973).

Berger's insight in this regard is remarkably instructive. Going well beyond the present question of the one and only symbolic universe, we see how force may serve totalitarianism by creating agencies of socialization that serve its legitimating interests, even if participation occurs under duress.

One need not go to the extreme of violence, however, to understand how mundane factors may lead to adjustments in a prevailing symbolic universe to the advantage of incursions of modernization. In the virtually certain event that the Third World society in question is not internally completely homogeneous, some interests will be better served by modernization than others (Berger, 1967). When representatives of advantaged interests join together to pursue their jointly shared preferences, they may persuade indigenous theoreticians that modifications in their symbolic universe are quite suitable. They may develop, moreover, an ideology to promote their interests.

A useful way of understanding ideology is to construe it as thinking in different ways within the same symbolic universe depending on material interests within a specific social system (Berger & Luckmann, 1966). This need not be as cynical and selfish as it sounds.

For example, migrant workers on the Trotsky family farm were afflicted with a temporary visual impairment, twilight blindness, due to a lack of vegetable fat in their diet (Trotsky, 1930/2007). Though the Trotsky family was far from wealthy, their migrant laborers were the poorest of the poor.

If they were to have sought the aid of a god of poor, afflicted, people of the land, it would have been perfectly understandable and hardly evidence of bitter self-absorption. The migrant farm workers would have been simply trying to survive. What better reason to invoke a God that served both their spiritual and material needs, especially in a nominally Christian nation such as late nineteenth-century Russia (Trotsky, 1930/2007).

While efforts such as this to modify or create a symbolic universe that acknowledges specific needs may fail miserably, insofar as they have a large number of like-minded, mutually supportive, reality affirming members, their chances of succeeding are enhanced. A community of committed believers, each attesting to the rectitude of the others, is an essential and compelling component of any prescription for success in gaining access to the symbolic universe, perhaps by winning the favor of universe maintaining officials (Durkheim, 1912/1995).

After all, who's to say that all theoreticians and guardians are of like mind? Some may be more open than others to the entreaties of social interests of a particular sort. Conflicting definitions of reality may give rise to conflicting social interests, and a once intact and unmistakably sacred symbolic universe may be fractioned, redefined, and otherwise weakened. There's nothing new about the triumph of the material over the spiritual, especially when onslaughts on legitimation come from a multiplicity of different directions.

It's worth repeating that while Berger has a longstanding concern with modernization, his interest in the sociology of religion is also quite compelling, and the two inevitably overlap (see, for example, Berger, 1967 & 1992). This, no doubt, is what has made him so instructively aware of the ways that spiritual and worldly domains intersect, providing valuable insights with regard to questions of legitimation and their basis.

Moreover, beyond legitimation, it is quite likely that those bringing modernization will, if need be, have access to mechanisms of physical force far superior to what is available to those who might oppose them. One way or another, pretty well without regard to legitimation, though more easily with it, modernization in a variety of forms will rule the day.

8 Modes of production and cognitive styles

Following Berger's lead, we recall that human beings' express themselves in many ways. In the present instance, using *The Homeless Mind* (Berger, Berger & Kellner, 1973) as our point of departure, think first of the nature of modern technology and the ways in which, during the decade of the 1970s, it was employed in generating economic growth. The organizations that brought technology to bear were structured hierarchically. This fact was abundantly evident to organizational participants as well those on the outside, and hierarchy became a pattern that was taken-for-granted by people who lived in a modernized social system (Berger [Ed.], 1990).

With hierarchy, moreover, comes anonymity. Some relationships are face to face, but the greater the hierarchical distance between participants, the greater the degree of anonymity. Rather than being a feature that predictably arouses nothing but envy, suspicion, or enmity, anonymity, too, may acquire a taken-for-granted nature as something intrinsically essential to the way a modernized society functions (Berger, Berger & Kellner, 1973).

Linking hierarchy and anonymity

At some point we might expect Berger to tie together the concepts of hierarchy and anonymity in discussing class membership. It is difficult to imagine that the vertical distance between positions was unrelated to occupational prestige and, especially, monetary compensation for time on the job and work completed (Berger, 1963 & 1986). However, this is an instance in which the seemingly obligatory invocation of social class, one that follows directly from the character of the organization of technology-intensive production, did not occur.

Certainly, none of what has been said so far should be taken to mean that hierarchy and anonymity were accepted with equanimity by all, and that they posed no problems for those who endured them. If hierarchy and anonymity had been suitably coupled with the concept social class this would have been clearer. For now, however, I am merely saying that hierarchy and anonymity were ubiquitous and accepted with remarkably limited destructive resistance. We will discuss this at greater length in Chapter 10.

Style of labor

Then and now, whatever the worker's specific job, one presumes he knows it reasonably well. As such, given the job's location in a technology-intensive organization, it imbues him with a style of labor that makes it relatively easy to move into another position of comparable complexity, dexterity, and intellectual demands (Berger, Berger & Kellner, 1973).

Whether or not the worker wants to move is not something that decision-makers socialized into the ethos of contemporary scientific management are inclined to notice. After all, labor is still a factor of production.

The most important aspects of the style of labor intrinsic to a technology-intensive organization much as that just outlined are not difficult to discern. Nevertheless, Berger's facility in describing the modernized world of work places a heavy burden on his usually sure-handed, straightforward prose. Berger is simply not at his best when recounting the organizational nature of modernized labor. Needless use of clumsy neologisms is a difficulty we've already introduced.

Mechanisticity

Mechanisticity is an ugly neologism if there ever was one (Berger, Berger & Kellner, 1973). This term means simply that organizational participants become imbued with the reality that they are productively linked to a machine-like process. Participants are aware, even if only implicitly, that they are integral parts of a rationally organized sequence of production. All this becomes a taken-for-granted part of the worker's way of thinking, his concrete expressivity, as do concomitants of mechanisticity, namely reproducibility and measurement. The worker is a machine part, easy to replace and singularly lacking in uniqueness (Berger, Berger & Kellner, 1973).

In other words, any set of workers could be readily supplanted by others with a modicum of comparable training; their performance can be effectively reproduced, and their activities can be measured and quantified. Quantification is often used and misused in an effort to determine how well workers are performing in tasks they have been assigned.

It seems odd that this should have been presented as something timely, and in need of multisyllabic neologisms. There's certainly nothing here that Marx didn't describe or foresee 150 years ago. Insofar as men physically and cognitively function as machine parts, Marx understood this in his earliest work, and he expressed this without off-putting neologisms (Marx, 1844/1988).

Componentiality

Much the same is true of componentiality, nearly as awkward a neologism as mechanisticity, though it's a bit harder to understand, both as a word that sounds like what it is describing and as a technical concept (Berger, Berger &

Kellner, 1973). As Berger sees it, the cognitive style of the modern worker betokens awareness that anything he participates in producing can be broken down into discrete parts or components.

The components, moreover, could be used in making a variety of products, with specific outputs depending on having the right components in the right sequence. Thus, everything produced in the technology-intensive workplace can be analyzed into discrete component parts. Similarly, everything can be disassembled and, if we remember the sequence of interdependence of the components, reassembled in exactly the same form as before. The same applies, moreover, to the assembling and disassembling of collections of workers.

Human beings as machine parts

It seems impossible to read this account of technology-intensive production and cognitive styles without surrendering to the judgment that human beings have themselves been reduced to the status of component parts. In that sense they are active and intrinsic participants in componentiality. It's also seems impossible to read this account without acknowledging Marx's foresight (Marx, 1844/1988). Mechanisticity and componentiality are but two manifestations of human being's expressivity whereby their activities are objectified as fully functional working components of a machine (Horkheimer & Adorno, 1947/2007).

That there is really nothing new about this is something Berger eventually acknowledges, and he clearly understands that workers are fully aware of their machine-part status (Berger, Berger & Kellner, 1973). Berger's assessment, however, is not dismissive or harshly judgmental. Instead, like a good value-neutral sociologist, he presents the most accurate rendering he can of things as they are, and he leaves value judgments to the reader.

Bear in mind, moreover, that we may be working with very complex outcomes, with every participant in the harmonious assembly of components responsible for only a very small part of the final product. Given that they know their specific jobs, there is no purely technical need for participants to be aware of the finished product, what it looks like, or what it does (Berger, Berger & Kellner, 1973).

In this way, means and ends become separated. The only thing that ties the variously located participants together is an "abstract logic of production" whereby they know what to do with the components peculiar to their task, and they could easily and with minimal retraining perform another task in the same overall production process (Berger, Berger & Kellner, 1973).

Berger's use of the notion "abstract logic of production" is, unfortunately, a bit vague. I don't doubt that the phenomenon exists and is genuinely efficacious in the productive processes to which Berger refers. Nevertheless, his treatment of this idea could be substantially improved, making it clearer and easier for the reader to grasp and see the reason for invoking it. Berger's use of the term "explicit abstraction" with regard to the functioning of bureaucratic organizations, as we shall see, is more accessible and descriptively pertinent, while serving a very similar connective-tissue purpose for a different kind of organization (Berger, Berger & Kellner, 1973).

Furthermore, as Berger is aware, the absence of a functional or technical need for workers to be familiar with the final product of their mechanistically linked efforts does not mean that there is no innately human need for understanding what has been done. One of several meanings of alienation due to Marx is alienation of the worker from what he produces (Marx, 1844/1988). While Berger acknowledges Marx's conceptual contribution in the form of "alienation," he oversimplifies it and fails to see its multifaceted nature. In this instance, however, it's clear that Berger is talking about alienation as meaning human beings denied access to the products of their expressivity. They are reduced to the status of producers of objectifications that they can never fully appreciate (Horkheimer, 1947/2013).

I don't know if that conclusion is part of an objective analysis or a value judgment. The distinction in this instance is difficult to make. Either way, it's an accurate characterization of the worker's position. Perhaps the value judgment is best withheld until we ask questions such as "is it worth it?" That's the way we will proceed from here on.

Consciousness and technology in a modernized world

The peculiar character of the processes just described marks the distinctiveness of institutionalized technology in a modernized world. By its very nature, we see that these organizational forms and the activities to which they give rise create an environment that is distinct and separate from participants' private lives. As we shall see, however, in a thoroughly modernized world it would be a mistake to exaggerate the degree of separation, an observation about which Berger sometimes seems both ambiguous and ambivalent. Specifically, does this separation give rise to pluralization as Berger uses that term, or is this something less consequential (Berger, Berger & Kellner, 1973)?

It is important to emphasize that the consciousness peculiar to a modernized technology-intensive process of production is the result of active participation in a distinctive physical environment. Thus, my inclination to characterize Berger's work as that of a philosophical materialist, a designation I'm sure he would reject.

Nevertheless, the style of consciousness that enables participants to do their work effectively and with a suitable kind and degree of intelligence is not something they bring to bear innately, but that is acquired through purposeful engagement. That this style of consciousness cannot be left at the workplace door to be picked up and used again the next day is self-evident. It's something that organizational participants inevitably, in whole or part, take with them as internalized (Berger, 1963).

9 Provisionally taking stock: "modes of production and cognitive styles"

At this point, one might stop and ask what Berger has accomplished so far that is of relevance to the sociologist interested in the relationship between distinctive institutionalized settings and participant's knowledge. As we have seen, this is the fundamental concern of a sociologist such as Berger who wants to know how people think and what they know. However, given that he has done little more than provide his own take on what makes the modernized world, and insist that comparable factors become intrinsic to our consciousness, one might conclude not very much.

After all, his presentation is not so compelling that it is impervious to wholesale disagreement. It's Berger's construction and it's made with the finite resources and the powerful but imperfect intelligence belonging to him. As such, we may find it unexceptional. This, one might reasonably conclude, is work done by an inveterate and willfully undisciplined sociological tourist, someone with his own labor-intensive way of doing research, but perhaps without the methodological rigor that most sociologists insist must characterize their activities.

Nevertheless, we should bear in mind that Berger is an accomplished sociologist of knowledge. His methods are his own, and over the years they have served him well. As a sociologist of knowledge, he is less interested in the institutional structures that manifest modernization than in their relationships to the way we think, to our style of consciousness. In light of the spillover from workplace to private life, the style of consciousness that is unique to a modernized world of work is of special interest (Auster, 1996).

For example, whatever the concrete details of the specific job done by participants in large, technology-intensive organizations, they must, as we have explained, develop a taken-for-granted understanding of hierarchy and anonymity in everyday social life. Furthermore, participants and their coworkers are subject to the mechanistic logic of componentiality that characterizes the overall production process. The overall production process organizes people according to the same logic that it organizes and assembles component parts. Organization is hierarchical, everyone is replaceable, and transitory social relationships acquire a machine-like character (Berger, Berger & Kellner, 1973).

Functionaries relating to functionaries yields anonymity

To make matters more complex, participants must unreflectively understand that those with whom they may have face-to-face, even intimately personal relationships are kindred functionaries, filling a role that could just as readily be filled by many others. Consequently, participants' relationships, even with those personally closest to them, must include an anonymous connection to the functional role occupied by the other. Participants unself-consciously develop the ability to accept the dualism that this kind of organization imposes, and to a greater or lesser degree relationships outside the workplace come to manifest the same social ambivalence along with the expectation that relationships will not be lasting (Berger, Berger & Kellner, 1973).

In this regard, it's worth noting that all of us play many roles over the course of a lifetime (Fine & Shulman, 2002). Each role may be so distinctive that it requires a separate and specific perspective. In a socially constructed world the institutions that give rise to these roles are inherently precarious. This is especially true under conditions of modernization when rapid social change may redefine, discard, or create roles in line with changing technological developments. And in truth, the coupling of intimate interpersonal relationships with roles that betoken anonymity is a distinctive feature of social systems characterized by modernization and modernity.

Anonymous social relationships were well known to Durkheim, especially as they occurred in urban environments where people routinely made incidental contact with many more others than in isolated rural areas (Durkheim, 1897/1997). In traditional communities, however, the limited number of contacts was much more likely to be with those with whom well-developed relationships had been established (Tönnies, 1887/2001).

However, the distinctive social character of well-developed relationships in the modernized world, as Berger describes it, is that they are commonly coupled with anonymity. The roles, moreover, may be the kinds that signify people coming and going and relocating as circumstances and the logic of production dictate. However, when actually playing a role, the activity calls forth the feelings that one would expect (Mead, 1934/1967). At least for the time being, the role with which one has been bestowed is real enough, even though participants may know that it may soon be superseded by another (Berger & Kellner, 1981).

Social roles and secondary socialization

Berger's formulation of the taking of social roles is, for the most part, conventional but interesting: Roles are socially bestowed, socially maintained through active participation, and they are socially transformed when processes of social change dictate such transformations. However, for the time being, the here and now, sharp departures from largely unself-consciously motivated, commonsense expectations for suitable role-specific behavior may be regarded as instances of pathology or deviance, and dealt with as threats to the institutions that the roles contribute to constituting (Berger, 1963).

Roles, moreover, are central constituents of one's identity. The more we play them with greater intensity the more we are so defined. This formulation strikes me as indisputable, and it makes me extremely uneasy with Berger's claim in *Invitation to Sociology* that, in effect, we can step in and out of roles with relative ease (Berger, 1963). Yes, this is true of roles which entail a level and kind of secondary socialization that lacks depth and intensity, such as the roles played by our baseball players or those in technology-intensive organizations. However, even roles that are due to secondary rather than primary socialization can have extremely strong and directive, unself-consciously motivated holding power. Following Berger in *Sociology Reinterpreted* (Berger & Kellner, 1981) with regard to secondary socialization into social roles:

> Socialization can be seen as an enormously powerful process whereby the "objective" structures of society "out there" are internalized within consciousness (Berger & Kellner, 1981: 92).

As constituents of man-made social structures, namely institutions, social roles and patterns of role relations are always in a precarious position, some more than others, depending on the kind of institution involved. As an important case in point, institutions that have been endowed with cosmic significance such as that which suffuses religious life may be construed as grounded in the universe – the eternal cosmos – rendering them powerfully resistant to the wear, tear, and chaos that afflicts secular, mundane social constructs. Some humanly constructed roles, thus, are given sacred status (Berger, 1967). Furthermore, the institutions in which they are embedded have substantially greater holding power than those that have not been endowed with nominally sacred status.

In the course of discussions of the objective facticity of institutions and roles such as the foregoing, I became a full-blown philosophical realist, affirming the unambiguous reality of sociological concepts such as institution and role. They are not merely for-the-time-being conveniences that minimize the burden of everyday decision-making and direct us to one set of suitable choices rather than another. They have an undeniable reality peculiarly their own, sui generis (Berger, 1967). Thus we see, indeed, that some institutions and social roles that are products of secondary socialization may have enormous, largely unself-consciously motivated holding power.

Given very different circumstances, roles acquire the dualism that comes with the functionary's anonymity noted above. Individuals experience themselves as having this same dual character, concretely unique and abstractly anonymous, existing as both specifically peculiar people and as replaceable component parts. Such a bifurcated existence is potentially troublesome, and imposes demands that the concretely unique self be given opportunities for development and expression lest excessive repression produce an emotional cripple unable to function as a component part or as a person (Berger, Berger & Kellner, 1973).

Social roles, bifurcation, and dualism

The bifurcation and dualism required by the anonymity we use in dealing with even our best-friend functionaries may be something new devised by Berger to aid his understanding of social relationships. In *The Social Construction of Reality*, Berger was quite emphatic: "In the face-to-face situation the other is fully real. This reality is part of the overall reality of everyday life, and, as such, massive and compelling" (Berger & Luckmann, 1966: 29). It is worth adding that "All other forms of relating to the other are, in varying degrees, 'remote'" (Berger & Luckmann, 1966: 29).

Berger's introduction of social roles characterized by dualism due to the anonymity required by modernization is, I think, more important than he acknowledges. The nature of persons and of social relationships has been insightfully redefined, for at least some purposes, in a way that I think may be legitimately characterized as extraordinarily creative.

Research on modernization, in this instance, has had an important effect on social role theory. One can't rework a discipline's basic concepts without deserving a good deal more attention than Berger has provided. Perhaps he's being modest.

In any case, Berger's notion of dualism due to anonymity is qualitatively different from his invocation of the notion of anonymity in reference to filling in typifications of people we don't know very well and are for the first time encountering face-to-face (Berger & Luckmann, 1966: 31–34). This is an entirely different process. It seems quite unlikely that Berger would have confused the fundamental transformation wrought by bifurcation and dualism with this less complex activity that is in no way essential to a revised understanding of the world or to transforming what is real.

In fact, Berger makes clear that in the presence of face-to-face interaction, the anonymity of typifications is always in the process of threatening to break down. The anonymity of typifications gives way to the concreteness of the individual. Typifications yield to characteristics of the real world, as we no longer rely so heavily on shorthand collections of traits that we commonly use in place of the richly detailed and fully substantial nature of things as they actually exist.

Berger and Mead

Berger's discussion of the complexity and richness of face-to-face interaction would not be possible without George Herbert Mead's fully developed account of reciprocally taking the role of the other (Mead, 1934/1967). Mead emphasized, moreover, that none of this would be possible without the power and versatility of the medium of language, a lesson that Berger has learned well, and with regard to which he has a great deal to teach.

Human beings' expressivity enables them to produce language in the form of objectifications. Language is construed as an objectification because it can communicate meanings that are not limited to inarticulate, ill-formed, transient utterances bespeaking merely a passing emotional state. Language is a coherent sign system, and as such permits the transformation of experience into a generally available object of knowledge (Berger & Luckmann, 1966: 69). When I encounter language it has a facticity external to me, and in various ways its effect is prompting and imperative.

The coherence of language and its existence independent of primitive vocal cords vibrating hysterically in a transitory, ineffectual expression of pain, pleasure, anger, or frustration give it enormous power:

> Language is capable of becoming the objective repository of a vast accumulation of meaning and experience, which it can than preserve in time and transmit to following generations … I can speak about innumerable matters that are not present at all in the face to face situation, including matters I never have and never will experience directly … Because of its capacity to transcend the "here and now," language bridges different zones within the reality of everyday life and integrates them into a meaningful whole (Berger & Luckmann, 1966: 37 & 39).

Given Mead's (1934/1967) contribution in the form of his appreciation of the indispensable place of language in social interaction, Berger's notion that the social reality of everyday life becomes less anonymous as we move from typifications most lacking in concrete detail to face-to-face encounters makes a good deal of sense. However, I have a lot of difficulty with Berger's claim that social structure is, in fact, the sum total of typifications and actual encounters along with the patterns of interaction they may create (Berger & Luckmann, 1966). This conception reduces the idea of social structure to the level of the individual or to an increasingly abstract surrogate or type.

Obvious structural factors such as prevailing interest rates, the shape of the occupational distribution, the distribution of wealth and income, patterns of social mobility, investment arrangements of large corporations, the internationalization of capital, and a broad range of other factors that belong in any account of social structure have not been acknowledged (Blau, 1975). I see the logic of Berger's position, but he misses too much, a good deal of which is of overriding importance.

Nevertheless, in understanding the "taken-for-granted," a commonsense world that we share with others as we progress through our daily routine and render it readily interpretable, Berger's formulation is compelling:

> The social reality of everyday life is thus apprehended in a continuum of typifications, which are progressively anonymous as they are removed from the "here and now" of face-to-face situation (Berger & Luckmann, 1966: 33).

This is so skillfully rendered that it's difficult to reject even when there may be good reason to do so.

Stock taken

Having taken stock of what Berger has rendered, we can respond to the question "What has Berger accomplished with regard to explaining modes of production and cognitive styles?" by acknowledging that, in contrast to our initial assessment, "quite a bit." Much of what he has written is of general theoretical interest and not limited to modernization and modernity. This is especially the case with regard to the innovative use of basic sociological concepts. His work so far, in spite of sections in which he presents established material as if it were something new has been theoretically compelling.

10 Modes of production and cognitive styles amplified

Spillover from the workplace

Again, given Berger's primary concerns, one of the most interesting and important aspects of this style of consciousness is that it is not tightly contained in the workplace. Instead it spills over into the private sphere, and even when not at work, participants may think and feel much as they do when they are on the job. In effect, the cognitive style of the technology-intensive world of toil and production becomes the cognitive style of the modernized world in all its forms (Burawoy, 1992; Willis, 1981).

This makes clear the experiential priority of the workplace in the modernized world. After all, both the world of work and the private world are socially created much as explained in *The Social Construction of Reality* (Berger & Luckmann, 1966). It is evident, therefore, that insofar as the private world is created by the conscious activity of workplace participants, they are imbuing it with characteristics that otherwise would be peculiar to the world of work.

Notice that Berger has not specified the kind of society, capitalist, socialist, or something else, in which all this is taking place. One can only surmise that as far as Berger is concerned, it makes little or no difference. The consequences of doing technology-intensive work in a modernized workplace will result in creation of people with the same or similar characteristics formed by working in the same kinds of institutional environments. The key factor is not the type of political system or its prevailing ideology, but the process of modernization and the nature of modernity (Berger & Huntington, 2002).

As far as I can tell, the presentation of consequences of doing technology-intensive labor are constant through the later years in Berger's career when he becomes a cheerleader for capitalism in promoting modernization (Berger, 1986, 1990, 2011 & 2014). Early on Berger puts it succinctly in the following quotation: "It is not correct to say that each society gets the men it deserves. Rather each society produces the men it needs" (Berger, 1963: 110).

This is an especially important idea because the world as it was organized in 1973 when *The Homeless Mind* (Berger, Berger & Kellner, 1973) was published has been undergoing a process of reorganization for nearly half a century. Insofar as the prevailing mode of production has been subject to substantial change, it will need and produce a correspondingly different kind of experientially determined men. In this sense, *The Homeless Mind* (Berger, Berger & Kellner, 1973) is in some respects obsolete. Nevertheless, the fundamental social processes at work, the ones that interest us most, are invariant from one setting to another. The relationship between institutions and cognitive styles is primary.

The place of the Protestant Ethic in this process is, I think, a matter of conjecture. To hear Berger tell it, the Protestant Ethic or something that has the same effect is to be found in the most unlikely places with consequences that promote capitalist development and high levels of productivity. Nevertheless, after repeatedly celebrating this cultural form, Berger continues to deny that he is a cultural determinist (Berger, 2011).

The sociology of knowledge

For now, at least with regard to the creation of styles of consciousness in a modernized world, Berger's work is compelling. Berger's contribution to the sociology of knowledge is quite real, and if we divest ourselves of the protective relevance structure of the sociologist, his analysis may be judged as troubling. He has clearly shown us that "the sociology of knowledge must concern itself with everything that passes for 'knowledge' in a society" (Berger & Luckmann, 1966: 15). And much of it is dehumanizing.

The notion that we are what we do is commonplace, but when we see the overarching consequences of participating in technology-intensive, modernized society, we may feel, to exaggerate a bit, that entirely too many people have been reduced to the status of economically productive automatons. What else is in their heads, save a vague but painful need for wholeness and fulfillment (Durkheim, 1897/1997)?

Odd as it sounds, I sometimes entertain the peculiar idea that Berger is a Weberian who is badly in need of a Durkheimian complement. Perhaps this is just a product of my own theoretical knowledge and interests. Nevertheless, understanding the modernized world as rendered by Berger seems badly in need of application of the best known concepts from Durkheim's early work, especially anomie, egoism, and organic solidarity. Marx's multifaceted concept alienation is also quite useful (Durkheim, 1897/1997; Marx, 1844/1988).

Perhaps all this makes it seem as if the modernizing and modernized worlds are heavy-laden with social pathology. In truth, if they're not, the potential for pathology is hard to miss when examining societies of this kind. Berger seems to be aware of this, but he also seems inclined to view the difficulties of modernization as easy to exaggerate and ultimately worth the pain and suffering (Berger, 1986). If we err in the opposite direction and exaggerate the difficulties, moreover, we may place ourselves in danger of romanticizing the past. But anyone who has lived in a rural area that closely resembles the circumstances or our forbearers will immediately recognize how cold, hard, unforgiving, and insular it can be.

When Berger asserts that “deception and self-deception are at the very heart of social reality” I want to ask him just why he bothered to write the first five chapters of *Invitation to Sociology* (1963). I learned long ago that explanatory science is impossible without determinism. Berger’s misguided efforts to debunk all of sociology in the concluding chapters of *Invitation to Sociology* (Berger, 1963) stand in stark contrast to the truly fine first chapters of the same book.

Nevertheless, the flawed and incomplete nature of any empirical discipline seems to be emphatically true. If our behavior were in fact socially determined, completely and in every respect, sociology would not be able to discern this. Again, I laud Berger for acknowledging the need for some sort of explanation for his insistence on free will, and for making the best argument he had. Even though it’s very brief, Berger’s explanation is the most credible one I’ve seen by a sociologist.

11 Reality as a social creation

When we acknowledge that Berger has convincingly demonstrated that the world is socially constructed, that it could be otherwise, the temptation to cast all morals as relative becomes quite forceful. Morals can be explained sociologically, but there is no sociological basis for making final decisions among different moral codes. Thus, we might ask, "Once one recognizes the 'constructedness' of moralities, how can one make moral judgments" (Berger & Kellner, 1981: 78)? The best answer I can come up with is "in the usual way," but not through reference to sociology as the domain of the final arbiter.

Modernization and trade-offs

If modern methods of production generate alienation, anomie, and egoism in the social world that they create, evaluation of the acceptability of these outcomes is the peculiar province of each individual. It should be clear by now that the sociologist's concern is not with evaluating the consequences of what people do, but with the action itself. This need not mean that we construe alienation, anomie, and egoism as outcomes that are not the sociologist's concern. But it does mean that sociologists leave for others questions as to whether or not modernization and modernity are worth the grief (Berger, Berger & Kellner, 1973).

In a sense, the development of sociology requires a rudimentary catalogue of human activity that is devoid of value judgments (cf. Berger, 1963). Again, it is one thing to say that alienation, anomie, and egoism are outcomes of modern technology-intensive production, and quite another to say they are acceptable or unacceptable trade-offs for the affluence made possible by enhanced productive power.

It is clear, moreover, that Berger's way of explaining how the modernized world is created through a process of social construction is a key constituent of his work. As we suggested earlier, it is no accident that *The Homeless Mind* (Berger, Berger & Kellner, 1973) and much of the rest of Berger's prolific output, deals with modernization and modernity, much as we have been discussing.

These are primarily processes that serve to illustrate Berger's way of doing sociology. In addition, however, they are among his primary substantive concerns. One who would come to terms with Berger's sociological perspective needs to firmly grasp his understanding of modernization and modernity and, especially, attendant theoretical developments. We have made a useful start on this endeavor in what has gone before.

As with his failure to clearly define the concept modernity, however, Berger is not always as helpful as he might be. Sometimes I get the impression the he wants us to fill in empty spaces with "Oh, Hell! You know what I mean." If so, sometimes he's right, but Berger has an annoying tendency to assume that we know things we don't know, or that we just can't be sure of.

Globalization

It is true, moreover, that thoroughly modernized economies such as that of the United States have undergone enormously substantial change since *The Homeless Mind* (Berger, Berger & Kellner, 1973) was published. Technological developments and the internationalization of capital have dramatically modified the modern world of work, in large measure by reducing or relocating the availability of conventional types of toil (Wilson, 1997). I think, however, that many of the most conspicuous and important characteristics identified by Berger remain roughly applicable in describing the everyday experience of most people today.

Nevertheless, the world of ordinary people is a good deal more difficult and the economy has been modified, as just noted, by transformative technological developments, and by globalization that, contrary to Berger (1986), has been accomplished largely in the interests of the very favored few (Stiglitz, 2003). But in matters of importance to our discussion, things are much the same (Habermas, 2015). After all, even computers are manufactured on assembly lines.

Furthermore, for Berger the Internet, to which he gives astonishingly little attention, is subsumed by globalization. Recall, moreover, that Berger's view is that globalization has been going on in various forms for over a hundred years. The Internet powerfully facilitates this process, but it is neither unique not absolutely necessary. One world-changing outcome of globalization is that the kind of work built into the foundation of *The Homeless Mind* (Berger, Berger & Kellner, 1973) is still being done but more of it is being done by the well-educated work forces of places like India and Pakistan (Bowles, Edwards & Roosevelt, 2005).

The more the workers in these culturally varied and distinctive countries deal with the kinds of work imported from the West, the more westernized they become. This may not represent validation of the convergence hypothesis, but it makes formulation of that hypothesis easier to understand (Tinbergen, 1956).

At least as early as publication of *The Capitalist Revolution* (Berger, 1986) Berger acknowledged awareness of globalization or the world capitalist system. While he did not develop this idea, nothing that he said suggested that he regarded it as inconsistent with widespread prosperity. One can acknowledge globalization as Berger understands it as an integral part of the process of modernization and, on balance and for the long term, decide that it works for the better of all of us, though this conclusion, to the present, has proved erroneous (Berger & Huntington, 2002). The consequences of globalization as discerned by Berger have the character of the wishful thinking of a hard-core capitalist. Nevertheless, for good or evil, the fact that almost all societies today exhibit the most telling characteristics of pluralism is attributable to globalization. Globalization fosters the kind of religious and other forms of diversity by posing challenges to existing plausibility structures (Berger, 2014).

In *Many Globalizations* (Berger & Huntington, 2002), Berger also acknowledges that cultural globalization has more sources than America or, more generally, the West. Instead, though this is another interesting idea that he leaves undeveloped, Berger posits the existence of alternative globalizations, meaning alternative modernization processes, and, perhaps most important, alternative modernities. Given Berger's unduly broad-brush, out-of-focus definition of modernity, however, the existence of alternative modernities doesn't really tell us much (Berger, 2014: 5).

Social construction and post-modernism

Nevertheless, it is very encouraging that Berger makes a point of separating himself and his work from the destructively excessive interpretation of social construction employed by radical post-modernists (see, for example, Zizek, 2011; Amin, 2014; Sloterdijk, 2013). Perhaps it shouldn't be necessary to offer the unequivocal judgment that there are, indeed, facts, and that some things are correct and others are not, but given the extremes to which some post-modernist authors have pushed the idea of social construction Berger's more conventional realism is a welcome development, indeed. When Berger observes that reality is "a quality appertaining to phenomena that we recognize as having a being independent of our own volition" (Berger & Luckmann, 1966: 1), he happily parts company with some of the more extreme post-modernists (see, for example, Deleuze & Guattari, 2009). Berger's term for dangerously lawless freedom, "anarchic relativism," fits post-modernism like a comfortably tight glove (Berger [Ed.], 2009).

When post-modernists make provocative pronouncements such as "the law of gravity is a social convention," Berger, I'm sure, would regard it as absurd (see Sokal & Bricmont, 1999). I suppose we could sweat and squirm and conclude that since they said "the *law* of gravity is a social convention" rather than gravity itself is a social convention, we might find some sort of sense, however pointless, in their seemingly nonsensical claim. I think, however, that such efforts would rightly be dismissed by Berger as just so much obscurantism. Yes, all facts are subject to social interpretation, but that is very different from denying their reality.

I know of no post-modern thinker who would endorse the following from Berger:

> I apprehend the reality of everyday life as an ordered reality. Its phenomena are prearranged in patterns that seem to be independent of my apprehension of them and that impose themselves upon the latter. The reality of everyday life appears already objectified … constituted by an order of objects that have been designated as objects before my appearance on the scene (Berger & Luckmann, 1966: 22).

For a thoroughgoing post-modernist, the overwhelmingly untenable claim occurs in the last sentence: The reality of everyday life not only appears to be "constituted by an order of objects," that actually exists as such before anyone sees it. A rock is a rock, a tree is a tree … a gas chamber is a gas chamber whether or not someone is breaking it, chopping it down … or dying in it (Sokal & Bricmont, 1999). These commonplace assertions are counter-intuitively at odds with a post-modern perspective.

When post-modernists acknowledge the existence of narratives, which is not all that often, they insist that all narratives are in conflict, and the conflict cannot be resolved. One wonders, paraphrasing Berger, how the most orthodox of them would respond to something like "there is the pedophile's narrative and the violated child's narrative" (Berger [Ed.], 2009). I imagine they would acknowledge the shock value, but contend that shock value and conflict resolution are not the same.

Berger's more temperate perspective, however, leaves unanswered important questions that have been raised with much greater force in recent years. Issues such as the environmental consequences of ongoing technological development, as well as who benefits and who does not from the abundance that accompanies modernization and globalization (Berger, 1986).

However, these difficulties can be brushed aside, if only for the time being, as we further investigate how Berger understands the nature of modernization and modernity. Inevitably, this will include recognition of what we take to be additional errors and ill-advised omissions, including Berger's failure to adequately discuss the basic phenomenon of social class within the context provided by *The Homeless Mind* (Berger, Berger & Kellner, 1973), an organizational characteristic of first importance in any modernizing or modernized society (Berger, 1963).

Social class in *The Homeless Mind*

We are not writing a review of *The Homeless Mind* (Berger, Berger & Kellner, 1973) but, as already noted, it is difficult to understand how an observer as astute as Berger could write an entire book about the social processes and consequences of modernization during the 1970s but make only scant reference to the inevitable emergence of a class system. This, in spite of the fact that in his first book, *Invitation to Sociology* (1963: 82), Berger took the position that "There is good reason to believe that some of the fundamental characteristics of a class system … are to be found in all industrial and industrializing societies, including those operating under socialistic regimes."

Berger also notes that "The most important kind of stratification in contemporary Western society is the class system" (Berger, 1963: 79). He attributes introduction of the concept class into social science discourse to Marx, thereby giving Marx full credit for an idea that has been central to discussion of social stratification in capitalist and other societies, and has demonstrated its utility in countless venues (Berger, 1986).

Furthermore, our current reference point, *The Homeless Mind* (Berger, Berger & Kellner, 1973), was written when the United States still effectively harbored a de facto race-based caste system whereby blacks were even more effectively than today disenfranchised and shunted into enclaves of disadvantage, where they remained separate from whites. Berger's active condemnation of this caste system would seem to constitute a violation of his overriding commitment to a value-free sociology.

Berger, however, asserted that it was part of the sociologist's professional knowledge that race and claimed differences among races are social conventions, invidious distinctions without a biological foundation. As such, failure to condemn the American race-based caste system was an instance of bad faith, meaning that "a free act is interpreted as a necessary fate" (Berger, Berger & Kellner, 1973).

It seems to me, however, that Berger is picking and choosing in a fundamentally arbitrary manner the institutions he will condemn. The social creation of race is part of the social creation of reality. Certainly no one expects this reality to be the same for everyone or to imagine that it is constructed in the same way everywhere. But it seems hopelessly out of character for Berger to intimate that sociologists should engage in biological reductionism in determining what institutions to condemn and what institutions to celebrate (cf. Berger & Huntington, 2002).

In any event, Berger is no stranger to the ubiquitous character of social stratification. Indeed, in *Invitation to Sociology* (Berger, 1963) he makes it clear that class is inescapably consequential with regard to a broad range of outcomes from life expectancy, to taste in music, to marriage partner, to the books on one's shelf, assuming one has sufficient resources to be able to afford books and a shelf. After all, class systems often emerge from the even more constraining, confining, and determinative caste systems frequently found in the Third World (Cox, 1948).

It seems odd, indeed, that something as diffusely powerful in its consequences would not have an effect on the degree to which one feels "homeless" or would in other ways lack pertinence to a discussion of modernization and modernity, or social organization construed more generally. Berger's failure to emphatically acknowledge the emergence and reality of social class strikes me as a damaging oversight.

One might surmise that over the ten years between *Invitation to Sociology* (Berger, 1963) and *The Homeless Mind* (Berger, Berger & Kellner, 1973), Berger revised his conception of the all-pervasive, massively consequential importance of social class, thereby rendering it a good deal less important. Reasons for this, however, are not to be discerned. Besides, in *The Capitalist Revolution* (1986) Berger takes the position that in most of the world class has superseded all other forms of stratification.

This is one instance, a fairly rare one, in which Berger gives unalloyed complimentary credit to Marx for his theoretical work. Marx introduced class to social science as a specifically economic phenomenon, and it has become the measure of choice to gauge participants' distance and distribution in a wide variety of social science disciplines when studying social stratification in thoroughly modernized societies (Berger, 1963 & 1986).

Marx's interpretation of class enables us to make clear that earnings, wealth – money! – have superseded traditional measures, such as status, power, and occupation as the preferred method of gauging social distance between individuals and groups. This in spite of the fact that Blau and Duncan in their influential and still widely read 1967 classic *The American Occupational Structure* (Blau & Duncan, 1967), written only six years before *The Homeless Mind* (Berger, Berger & Kellner, 1973) held that the occupational distribution was the backbone of the social stratification system in the United States. Moreover, the few comments they made regarding Marx were patently derisive.

It may seem that we have returned to *The American Occupational Structure* (Blau & Duncan, 1967) too many times, exaggerating the merit and influence of one instance of a now badly dated piece of empirical research. However, Blau and Duncan's (1967) book still has many sophisticated and very important empirical and methodological lessons to teach us. I always found it difficult to illustrate the simultaneous occurrence of structural mobility and individual mobility until I studied the mobility tables that appear early in *The American Occupational Structure* (Blau & Duncan, 1967).

Furthermore, I couldn't think of a reason, other than ideologically engendered blindness, why anyone would claim that educational attainment was a more powerful determinant of status attainment than class until I examined Blau and Duncan's simple path model (Blau & Duncan, 1967). I'm sure that a similar model estimated today would give very different results, but Blau and Duncan's book was written during the prosperous and optimistic post-World War II era when prospects for the future seemed endlessly hopeful, especially for the well-educated (Newman, 1993 & 1999; Schwarz, 1998).

At the end of *The American Occupational Structure* (Blau & Duncan, 1967) the authors claim that ongoing abundance, increased and equitably shared, was guaranteed by the ascendance of the norm of universalism. The notion that a fragile, culturally pervasive, social psychological variable, the front end of one of Talcott Parson's pattern variables, the back end being particularism (Parsons, 1971), was the primary factor that undergirded the further development of a prosperous, meritocratic, and equitable society is, in my view, the most quaintly outmoded part of the book.

In 1967, however, no one except an occasional Marxist who doubted the continuing efficacy of Keynesian economic policy would have taken exception to Blau & Duncan's (1967) assertion (see, for example, Baran & Sweezy, 1966). It was a rare observer who foresaw the displacement of universalism by an American oligarchy and its attendant ethos of the reign of the billionaire (Ravitch, 2014). Recall, however, that it was Weber, writing 100 years ago, who understood attainments of members of the highest class as due to money rather than merit (Roth & Wittich, 1978).

Nevertheless, we sometimes find that intellectuals such as Berger who lived through the era of post-World War II prosperity have remained there, as manifest in their work (Berger, 1986; Berger, 2011; Berger & Huntington, 2002). In truth, it is difficult divest one's self of the meritocratic assumptions of continuing social and economic development that pervaded that time.

Furthermore, in many ways, *The American Occupational Structure* (Blau & Duncan, 1967) remains quite timely. It is a solid and innovatively crafted bench mark that provides an instructive basis for comparison with respect to recent developments. If the empirical results had remained valid and had come to encompass all ethnic groups, not only whites, legitimation of American society would not be in jeopardy.

Class is typically related to other measures of stratification, but it has the virtue of being much easier to measure, say in dollar terms, than alternatives. Furthermore, the related factors that come with class membership as measured by Marx, namely accidents of birth, circumstances of upbringing, and contingencies of the life course are less thoroughly confounded and confused with class than with other measures, such as status group membership or access to the implements of political power.

Berger's failure to do justice to social class as a phenomenon implicated in the process of modernization may be due to his contention that class systems are typically accompanied by a high degree of social mobility (Lipset, 1960). This is a vague statement, and Berger acknowledges that he is referring to a combination of upward and downward mobility. However, whether using Marx's ubiquitous characterization of class or invoking Weber's understanding of class as determining one's life chances, Berger makes the hard-nosed assertions that nothing can be done about the class structure, and that downward mobility for some is simply a fact of life (Berger, 1963).

It would be unfair to fault Berger for not being able to see thirty or forty years into the future. But if he had been given such powers of divination he would have done well to reconsider his position on social mobility, with movement in both directions and upward mobility holding a preponderant position (Newman, 1999).

Since publication of *The Capitalist Revolution* (1986), changes in the occupational distribution have sadly recast their pattern in the United States. Recent research suggests that cross-generational upward mobility in the

United States has stagnated, and perhaps the incidence of upward mobility has been surpassed by downward mobility (Putnam, 2015). Furthermore, the once sacrosanct relationship between educational achievement and economic upward mobility has become decidedly uncertain (Brown & Reay, [Eds.], 2015).

Nevertheless, discussions of this nature do not appear in *The Homeless Mind* (Berger, Berger & Kellner, 1973). Whatever changes have occurred over the decades, insofar as modernization brings class with it as a stratifying mechanism, I think it deserves much more attention than Berger has given it (Berger, Berger & Kellner, 1973). In my view, the emergence of a class system is, without question, properly construed as a manifestation of socially consequential diversity. As such, given Berger's intense concern with pluralism in the genesis of secularism, I am unable to understand why class has been given short shrift.

I suspect that part of my problem with Berger in the foregoing paragraph stems from the fact that the concept "pluralism" brings to mind a definition that, though closely related to Berger's, is peculiar to me. I construe pluralism to mean a surfeit of credible, often overlapping choices sufficiently persuasive to pose a compelling threat to existing plausibility structures. This is consistent with Berger's emphasis on the sacred/secular dichotomy, but it covers a good deal more. It places at risk commitments of all sorts, not just religion. Anything that might lend credibility to the claim that some things are certain, say parents love for their children, the belief that your nation is worth fighting for, or assurance that your spouse is traditionally faithful, and other socially pervasive sources of certitude are rendered doubtful.

I see no reason other than narrow and rigid disciplinary claims why pluralism should be limited to religious manifestations. After all, affirming the rectitude and stability of the examples just given seems rightly construed as taking a stand for the sacred and not surrendering to the secular, as those terms are commonly used by ordinary people.

Solidity of institutions and ephemeral meanings: a matter of emphasis

In *Invitation to Sociology* (Berger, 1963) Berger gives Durkheim a great deal of credit for apprising us of the truth that social facts exist *sui generis*, as objective and intransigently immovable as multi-ton rocks blocking the middle of the highway. In *Invitation to Sociology* (Berger, 1963), *The Social Construction of Reality* (Berger & Luckmann, 1966), and *The Sacred Canopy* (Berger, 1967), we learn that thoughts, meanings, and ideas, accumulated over time, give rise to structures endowed with objective facticity as they are made manifest in language and take shape in the form of roles and the institutions they constitute.

Moreover, institutions are things that cannot be thought away or willfully ignored, and they purposefully shape much of our behavior. We posed this question earlier: "How is it possible that subjective meanings become objective facticities" (Berger & Luckmann, 1966)? And we discussed it thoroughly. Subjective meanings and the objective facticities that constitute social structures are two sides of the same coin, complementary terms in a dialectical process involving externalization, objectivation, and internalization.

However, when Berger declares himself a Weberian, one for whom, as he sees it, thoughts in the form of human intentions produce a substantial part of the reality we share with others, though he's right, he seems to vastly diminish the importance of the intransigent reality of objectified social structures in favor of the meanings that may be their precursors, accompaniments, or constituents. It's all well and good to say "society defines us, but is in turn defined by us" (Berger 1963: 129). However, this in no way implies that the influence flows in some measurable fashion with equal force in both directions, especially after determinant structures have taken substance and shape.

We noted earlier that Berger uses the concept "objectivation," borrowed from Weber, to mean that what once existed only in consciousness is made manifest as external reality through the medium of human activity (Berger & Kellner, 1981). It is still not clear why Berger sometimes uses Weber's term objectivation and elsewhere employs the English word objectification, which, according to some English language dictionaries, has exactly the same meaning (*Oxford Learner's Dictionary*, 2015). Berger's unwillingness or inability to make up his mind presents us with what certainly appears to be another inconsistency, though one that is not particularly consequential.

Berger might irascibly respond to this by insisting that he long ago made up his mind with the knowledge that social structure is "built up by activity that expresses subjective meaning" (Berger & Luckmann, 1966) whatever we call the process whereby this occurs. He might add that we're making something out of nothing,

That observation may be true, and it's certainly informative, especially if we are willing and somehow able to trace social structure to its origins. This effort would follow the admonition that "the stuff out of which society and all its formations are made, is human meaning externalized into human activity" (Berger, 1967: 8). But what I really want to know is why, for present purposes, Berger gives priority to ephemeral meanings over the solidity of thoroughly established social structures.

At this point some might argue that the meanings I'm referring to are really not so ephemeral. Once they have been expressed in language, they share the ontological status, the objective facticity of social structures, because language enables us to transcend the here and now. Furthermore, objective social structures and structures of consciousness stand in dialectical relationship to each other, and the dialectic gives priority to neither element of the pair.

There is merit to these observations, but it remains the case that a meaning, say the meaning a baseball player gives to hitting a home run, is possible only because there is a complex and materially palpable institution called baseball. The difference between meanings and determinant social structures is self-evident.

The Protestant Ethic and the Spirit of Capitalism

I think it's likely that Berger's inclination to give priority to meanings over determinant structures may be due to the conspicuous place American sociology has given to Weber's book-length essay *The Protestant Ethic and the Spirit of Capitalism* (Weber, 1920/2010), a work consistent with the priority of meanings over modes of action and institutionalized organization. In truth, I found Weber's widely read and frequently referenced and celebrated essay surprisingly unconvincing, though I acknowledge that may be due to my lack of applicable historical knowledge. Nevertheless, the following from Berger seems quite pertinent:

> In Weber's view, it was Calvinism and its various offspring denominations ... that made a crucial contribution to the spirit of capitalism [a phrase that without violating Weber's intentions might be translated as the economic culture of capitalism]. The Calvinist of 'double predestination' holding that God has decided before time began who will be saved [the elect] and who will be damned, naturally led to pervasive anxiety concerning one's own fate in this frighteningly depressing scenario (Berger, 1986: 100).

As I read Weber it was clear that the central place in Calvinism clearly belongs only to the elect. Why the others, consigned to reprobation, continued to work hard and remain ascetic and abstemious to glorify a God who had capriciously condemned them to eternal damnation is very hard to understand (Stroup, 2009).

This is an obvious difficulty that Weber neither acknowledged nor explained, and the same applies to Berger (Tawney, 1926/2015). Perhaps it's accepted uncritically because it can be construed as a manifestation of the thoroughgoing rationalization of Calvinism – no miracles, no mysteries, and no uncertainty – and thereby consistent with Weber's thesis that rationalization was pervading the social world. However, I fail to see the rationality in making elect status something that is ascribed and cannot be earned. In my view, it's self-defeating and thoroughly irrational, unless rationality is rendered wholly sterile and totally devoid of substance.

Can we know the elect in the here and now? Perhaps we could if we adopted the assumption that those who do well – who are markedly prosperous and held in high esteem – are judged to be those whom God would not materially bless on earth if they were not to be spiritually blessed in the hereafter. I think this assumption is thoroughly dubious, but it makes a sort of wishful-thinking sense for the wealthy. Berger gave the endlessly hard-working, thoroughly ascetic, and joyless the status of the "other-worldly abstemious" to distinguish their approach to salvation from that of monks, religious hermits, and others who were "inner-worldly abstemious."

I must say that I've always found the practice of other-worldly abstemiousness odd in that it places the cart before the horse: Election precedes practice. I know my eternal destination because of the way I live now, promising election rather than effective denunciation. The practice bespeaks a fundamental terror unnoticed by Berger, Weber, and all others known to me. Weber's learned effort to counter Marx's account of the genesis of capitalism requires a real stretch.

Is this rational? Rationalization is manifest in Calvin's universe only if we construe it as having an essentially mathematical character. A place where everything is defined and quantified once and for all and no modifications may be considered. This is an abysmal cosmology for all but the elect, and it has no purpose that I can discern. Perhaps being devoid of purpose is a manifestation of rationality, kind of like playing bridge.

I think that Berger may have spent his entire professional life up to the present overvaluing *The Protestant Ethic and the Spirit of Capitalism*. If so, he's in good company (Parsons, 1937; Gerth & Mills, 1958; Roth & Wittich, 1978), but elsewhere Berger does such a compelling job of persuading us of the intransigent facticity of the social world that he, I presume unintentionally, undercuts his case.

An explanation does not occur to me, and transient meanings, for the short term, seem so weak when compared to already existing institutional structures that Berger's commitment to what I take to be Weber's position on meanings begs explanation. One wonders if Berger has conveniently invoked his dialectical process consisting of the three moments of externalization, objectivation, and internalization and forgotten the first two (Berger, 1967).

I'll take a big risk and offer the suggestion that Berger's fascination with Weber's focus on understanding the meanings that people entertain in their various places in a social system prevents him from seeing what I take to be the conspicuous position of social institutional structures in Weber's sociology. Society may be created by the meaningful acts of individuals, but the meanings have to come from somewhere, some sort of lived experience that inevitably takes place within established institutions (Berger, 1963).

Whether Weber was right or wrong about the emergence of capitalism in the West, it is also the case that anyone who reads translations of Weber's work will be struck by the frequency with which he presents explanations and arguments that read as if he had a lot more in common with Marx than Berger and most other sociologists seem to think possible. Perhaps this is additional evidence that Weber gave priority to structures of objective facticity as opposed to transient meanings. For example, take this quotation from *The Theory of Social and Economic Organization*:

> At an earlier period every worker could be said to have been primarily interested in becoming an independent small bourgeois, but the possibility of realizing this goal is becoming progressively smaller. From one generation to the next the most readily available path for advancement both for skilled and semi-skilled workers is thus into the class of technically trained individuals. In the most highly privileged class, at least over the

> period of one generation, it is becoming more and more true that money is overwhelmingly decisive (Henderson & Parsons, 1947: 425).

The preceding paragraph could have been written by Marx. It reads like a paraphrase of material taken from *The Economic and Philosophic Manuscripts of 1844* (Marx, 1844/1988). It clearly gives priority to social organization over member's meanings, effectively thwarting actions that they invoke. It also makes clear that the merit-based outcomes that ostensibly come with education, training, and experience do not apply to the upper classes, for whom money is the key to further gains, a set of plutocratic circumstances that seems clearly to apply today.

Berger's statement that "sociological understanding involves the interpretation of meanings present in society" is interesting but incomplete (Berger, 1963: 118). As we see throughout this document, social structures quite often exist unself-consciously, as in the case of thoroughly internalized social roles. Furthermore, some macro-structures may be man-made organizational phenomena that completely escape our awareness until we encounter them in a purposeful way (Berger & Kellner, 1981).

The following quotation from the same source as above is also instructive, as sharply coercive as it is brief:

> If the means of production and raw materials are under unified control, there is the possibility of subjecting labor to a stringent discipline and thereby controlling both the speed of work and standardization of products (Henderson & Parsons, 1947: 426).

A simple enough statement, but again one that could easily and with good reason have been written by Marx, and which unmistakably gives priority to social organization over meanings. It's hard to imagine that the meanings that may be shared by laborers in this context merit consideration in determining what is going on. They work, whether they like it or not, to make a living. Besides, the laborers' meanings are likely to be consistent with – because determined by – the work they do. Whatever their thoughts, they have no control over the nature of their tasks' level of compensation. If they don't understand this they are suffering from a failure of socialization.

Furthermore, those for whom the laborers work have only one concern, namely the rationally calculable pursuit of profit. These are exactly the same circumstances that prevail in a modernized capitalist society of today, and the same applies to the following from Weber's *Economy and Society*:

> For employers there is always the possibility of transforming their exploitation of labor into a mere source of income ... it also regularly tends to place workers in a state of dependence on people with whom they deal who enjoy a more favorable market position (Roth & Wittich, 1978: 129).

Again we see Weber writing about structured conflict, in this instance, fittingly, between capital and labor. Surely this statement is fraught with meanings, but the overriding reality in the situation described is made manifest is the objectification of social structure.

I realize that anyone can pick and choose affirmations suitable to his purpose from long and topically varied books. Nevertheless, I think the statements quoted above are typical of Weber and lead the reader to the conclusion that structure is the more conspicuous and determinative partner in the mutually implicative interaction between structure and meaning.

I think that in the unlikely event that Weber had been asked to assess the social problems that existed in an institutionally distinctive area, he would have proceeded much as Berger recommends: Investigate "not crime but the law, not divorce but marriage, not racial discrimination but socially defined stratification, not revolution but government" (Berger, 1963: 37).

My conclusion in the foregoing is speculative, of course, but I think it's consistent with what we know of Weber. In each instance, institutional structures are given priority over meaning-laden behaviors, though I recognize that the two are mutually implicative rather than mutually exclusive. Nevertheless, in drawing a distinction between social problems, as that idea is ordinarily understood, and sociological problems as the sociologist construes them, Berger gives priority to institutional structures.

He may very well find that the structures are thoroughly in need of debunking. In the process, however, he may discover an alternative, true-to-life set of structures that enable us to better understand how our world works.

I think that Berger's inclination to give priority to meanings seems more an aesthetic preference than an analytical judgment, one that he adopted when he was very young working on his dissertation. Whether or not I'm correct in this assessment, the farther one goes in reading Berger's work, especially that which came later, say after publication of *The Homeless Mind* (Berger, Berger & Kellner, 1973), the clearer it becomes that given his emphasis on meanings, Berger should have no reason to object to being characterized as a cultural determinist.

Nevertheless, he rejects this designation, asserting that he is not, as he puts it, a "culturalist" (Berger, 2011: 144). In this instance, I think the reader deserves a much better explanation than the very sketchy one that is given. After all, culture, in this context, consists of meanings, not the determinant structures that constitute institutions.

I have to admit, however, that my uncertain characterizations of Berger as a philosophical materialist and then as a cultural determinist leave me uneasy. No one likes to be unfairly labeled, but everyone still breathing has a point of view that others may try to interpret and name, often erroneously.

Still, in a concession to Berger's Weberian perspective, adequately characterizing the philosophical basis of his point of view can't be done without acknowledging his laudable commitment to the notion that material practices and social circumstances are reciprocally related to the cognitive content and style in which they are implicated. More than most other sociologists writing in the classical tradition, Berger explicitly demands that consciousness be given its due. As an empirically astute sociologist of knowledge, this, no doubt, is as it should be, and is consistent with his Weberian orientation, even if Berger overinterprets it – perhaps misinterprets it – by quite a bit.

Nevertheless, and to Berger's credit, he acknowledges that Weber's perspective harbors substantial limitations for contemporary scholars. Berger puts it succinctly as follows:

> I have never been much taken with the endless enterprise of interpreting and reinterpreting the Weberian opus. I'm convinced that Weber asked many of the crucial questions; I tend to think that many of his answers are no longer tenable and that today we must ask many questions that he could not possibly have asked (Berger, 1986: 7).

Berger might have added that textual exegesis has its place, but some Marxists and classical scholars, as well, have yet to acknowledge its limits.

The base/superstructure model and plausibility structures

Moreover, Berger is one who gives interesting, interpretable, and suitably brief meaning to Marx's base/superstructure model. Given his commitment to Weber, this is not something I had expected, but Berger's account is quite persuasive. In effect, the base is the locus of human activity, and the superstructure of the ideational and institutional products of that activity (Bottomore, 1956; Berger & Luckmann, 1966; cf. Poulantzas, 1978/2015). Once it's pointed out, the fact that the base/superstructure model is homologous with a plausibility structure is easy to see. The base/superstructure model is ideal in succinctly capturing the conceptual character of plausibility structure, an insightful idea whose provenance is due to Berger and for which he should be lauded. It also fits neatly into our discussion of the relative social impact of meanings and determinant social structures. Berger's interpretation enables the sociologist of knowledge to come to terms with the base/superstructure account in a very productive way.

This, I think, provides a useful perspective for thinking about the material, social, and cultural transformations wrought by modernization. It also helps us understand my commitment to the solidity of institutions and the comparatively insubstantial and ephemeral short-term character of meanings (Bottomore, 1956).

I have to admit, however, that I do not understand Berger's inclusion of the work of Nietzsche in any part of his discussion (cf. Berger & Luckmann, 1966: 7). This seems an awkward and misguided way to introduce the sacred/secular dichotomy. Surely Berger doesn't need to refer to Nietzsche just to borrow his provocative "death of God" proclamation. Perhaps this reflects my ignorance and deeply ingrained aversion to Nietzsche's work.

I had thought, moreover, that until Deleuze in 1966 first published *Nietzsche and Philosophy* (Deleuze & Tomlinson, 1983) my take on Nietzsche – difficult to take seriously as a philosopher, a social theorist, or a person – was more or less widely shared. George Santayana's characterization of Nietzsche as a constitutional invalid seemed as good a rendering as one could hope to find (Smith [Ed.], 1920/2008).

As it is, I can't see that Berger's work is in any way enhanced through reference to Nietzsche. The paragraph which he devotes to Nietzsche in *The Social Construction of Reality* (Berger & Luckmann, 1966: 7) is in no sense compelling or even perfunctorily informative. Berger's evocation of Nietzsche's "God is dead" pronouncement is hardly necessary to maintain his narrative as it appears in *A Far Glory* (Berger, 2002). Furthermore, beginning Berger's book *In Praise of Doubt* (Berger & Zijderveld, 2009) with reference to Nietzsche seems to me an invitation to misunderstanding.

One who would read all of Berger's best books might get the impression that Nietzsche has long been a lot more influential and informative than is actually the case. Surely Berger can find a better reference than Nietzsche. Nevertheless, I repeat that my knowledge of Nietzsche's work is abysmally deficient. Perhaps he has long been more important and persuasive than I had understood.

Even when we disagree with Berger, however, we should bear in mind that he is not a one-trick pony. Berger effectively juxtaposes quotations from Durkheim and Weber in a way that shows that he knows there is more to sociology than meanings. Following Durkheim, "The first and most fundamental rule is: Consider social facts as things" (Durkheim, 1898/1986). And from Weber: "Both for sociology … and for history, the objective of cognition is the subjective meaning-complex of action" (Berger & Luckmann, 1966: 18). These are hardly the comparative renditions of a doctrinaire ideologue of any sort.

For my own reasons, I place much heavier emphasis on Durkheim's argument for the facticity of social phenomena. Nevertheless, it is clear that Durkheim's and Weber's statements are complementary rather than contradictory, and sociology would be incomplete without both. I often find myself under-estimating Berger's appreciation of Durkheim, but the following quotation makes clear my mistake:

> It is precisely the dual character of society in terms of objective and subjective meaning the makes it a reality "sui generis," to use another key term of Durkheim's (Berger & Luckmann, 1966: 18).

It may be that when I raised a question as to Berger's comparative emphasis on the ephemeral character of meanings and the solidity of social structures, the intrinsic relationship between them inevitably generated confusion. I think, however, that the distinction and its importance are clear enough, and the same applies to the work of Weber.

Bureaucratic environment

We have already acknowledged that, for Berger, it seems that rational bureaucratic organization is an essential structural prerequisite to the effective functioning and sustained productivity of a modernized, technology-intensive economy. This may seem an unlikely position for a person on the political right, such as Berger, but bear in mind that economic organizations themselves have a bureaucratic apparatus, one that is closely geared to the process of technology-intensive production as we have described it (see, for example, Burawoy, 1992; Willis, 1981).

However, to maintain the social and political stability and support needed by a modernized economy, a massive political bureaucracy may also be essential. As far as I can tell, Berger uses the term "political bureaucracy" to reinforce the distinction between bureaucratic organizations that are tied to technology-intensive organizations of production, and those that create a stable and workable environment in which technology-intensive production can occur. This, however, is speculation as Berger offers no explanation. Moreover, as with the process of technology-intensive production itself, political bureaucracy both presupposes and creates ways of behaving and a cognitive style of its own (Berger, Berger & Kellner, 1973).

Furthermore, as we have already acknowledged, in *The Capitalist Revolution* Berger treats bureaucratization as a source of economic inefficiency, an important reason why socialist economies are comparatively weak (Berger, 1986). I have no doubt that this is something Berger suspected all along. In light of his discovery of capitalism, it may have come to occupy a more conspicuous place in his thinking about the social world and the essential but ideally circumscribed role of bureaucracy.

The cognitive style of political bureaucracy

For the most part, the cognitive style with which political bureaucracy imbues participants is compatible with the cognitive style inherent in technology-intensive production. Since both bureaucratic organizations and productive processes are first and foremost manifestations of thoroughgoing functional rationality, this compatibility comes as no surprise. Nevertheless, as we know, substantive rationality takes a variety of forms, so differences between the two kinds of organizations need not be evidence of malfunctioning or faulty design.

As with technology-intensive production, many aspects of bureaucratic organization are well understood by nearly all members of a modernized society. Activities such as how to approach bureaucracy, what to expect from it, and how to behave once inside are generally well known.

It is unlikely that anyone is formally taught how to use bureaucracy for ordinary purposes, but the information needed pervades a modern social system in good part, no doubt, because so many essential activities have been bureaucratized. In a real sense, participants' lives in a modernized society are surrounded by bureaucratic organizations, and citizens have more relationships for different purposes with political bureaucracies than with the workplace (Berger, Berger & Kellner, 1973). This is a set of circumstances pregnant with possibilities for informative research by a capable sociologist of knowledge.

As with modernized technology-intensive production, participation in a bureaucratic environment is anonymous. The peculiar characteristics of individuals who work in, or are clients of a bureaucracy are of no interest except insofar as they are suited to a specific social role that makes them eligible for bureaucratic affiliation. Berger uses Weber's term "affective neutrality" in making the case that legal-rational bureaucracies in a modernized society have what he terms a moral dimension (Berger, Berger & Kellner, 1973). The morality inherent in bureaucracy manifests itself in equal treatment for all; in other words, the affective neutrality just mentioned.

Whether or not this orientation to human beings, their characteristics, and their relationships is also inhumane is a judgment that falls outside the purview of the sociologist, as Berger understands his role. The sociologists' relevance structure enables him to acquire a great deal of information about bureaucratic organizations and the ways that people are categorized and moved about. Nevertheless, as we have seen, Berger holds that the same relevance structure prohibits value judgments that might find bureaucracies perniciously dehumanizing and offensively coercive in a passive-aggressive sort of way (Berger & Luckmann, 1966).

Still, as Berger sees it, in contrast to organizations geared to technology-intensive production, political bureaucracies have a morality of sorts that cannot be reduced to laws intrinsic to science-based technology and it's most efficient use. At the same time, however, bureaucracies also have the anonymous character discussed above, so Berger invokes yet another descriptively suggestive term, "moralized anonymity" (Berger, Berger & Kellner, 1973).

For the uninitiated, what Berger has to say about bureaucracies is illuminating. Bureaucracies produce nothing, but they are characterized by an all-consuming sense of orderliness. When this orderliness is linked to a set of categories, every member of the same category is ostensibly equal with regard to access to designations, statuses, and stipends that the bureaucracy may provide. Thus, for example, one is eligible for social security payments or one is not, depending on dispassionate adherence to pre-established rules. As already acknowledged, it is this categorical equivalence and strict adherence to pre-established rules that provides the basis for bureaucracy's morality (Berger, Berger & Kellner, 1973).

Bureaucratic decision-making, as a result, is predictable. Any departures from application of the rules that render it so may be construed as instances of corruption, perhaps exposing employees and clients to legal sanctions. In this sense, bureaucracies operate with the force of law. As a result, in the ideal, participants can expect a kind of justice from bureaucracies, the justice inherent in knowing that one will be treated as all others are treated, with no special privileges for nominally special people (Goodsell, 1983).

Berger's extended analysis of bureaucracy may seem gratuitous because so much of this material is well known (see, for example, March & Simon, 1958/1993; Gormley & Balla, 2012). But Berger takes it seriously, perhaps because Weber did, and because not all readers can be assumed to be equally knowledgeable with regard to important sets of issues, even when they pertain to knowledge that is widely shared. A century ago, when Weber was still alive and long before complex organizations had become a standard area of research and teaching for social scientists in a variety of disciplines, the material on bureaucracy would have been invaluable in a monograph on what then passed for modernization or the character of organizations in a modernizing or modernized society (Perrow, 2014).

Nevertheless, as far as I can tell, Berger's analysis is not original, and its reason for being is to contribute to understanding the nature of modernization and modernity, not organizations, as such. Perhaps if Berger acknowledged difficulties such as those briefly documented below, his efforts would seem a bit more interesting, a complement to what we take for granted rather than a straightforward repetition of what we already understand.

Bureaucratic muddling through

Anyone who has ever had the misfortune of being in a job that required writing procedures manuals for bureaucratic organizations will understand that the process of functionally rational bureaucratization is always imperfect, with significant departures from the ideal-type, something that Weber understood very well. There always seem to be exceptions to the rules in place, things that were not foreseen, sometimes sufficiently important to prompt revisions. The social world is infinitely complex, and bureaucracies are inevitably finite in their functions and responses.

For the most part, however, the rational, rule-governed framework envisioned for any bureaucracy is approximately attainable, but lingering uncertainty and organizational ambiguity is something that merits more attention than Berger gives it, perhaps because his material is purposefully presented at high level of abstraction. Berger's account of bureaucratic organization, informative though it is, remains unsullied by the grit, grime, and contortions of real-world transactions, while ironically, Weber's understanding of likely prospects for bureaucracy was more realistic (Andreski, 2006).

In contrast to technologically productive organizations, bureaucracies have no separation between means and ends. In fact, employment of proper procedures is as important as desired outcomes. Outcomes due to procedures that depart from those prescribed may very well be taken to be fundamentally flawed and worthless (Berger, Berger & Kellner, 1973).

In a world as complex as a modernized society, however, experience and common sense force us to acknowledge that a political bureaucracy, however carefully conceived, cannot foresee all pertinent eventualities. It becomes certain, then, that some functionaries will be faced with decisions that are not fully governed by the strictures in place. Events such as this will either prompt innovations – departures from strict adherence to prevailing rules – or at least part of the bureaucracy's operation will come to a halt, thwarted by the inevitable incompleteness of organizational design and the understandable unwillingness of bureaucrats to take risks by initiating innovations (Goodsell, 2014).

I have seen these difficulties first hand and was given the task of dispelling them. When all is said and done, for the long term and a complete sequence of events, it is simply not possible. I proceeded to willfully ignore this for some time, but had I stayed in the same job I would have eventually been stymied, forced to rewrite bureaucratic rules that were ostensibly precise and permanent.

Once again, I think that Berger's own exaggeration of Weberian purism, something that Weber himself did not embrace, prevents him from understanding how important parts of the modernized world actually work. We can overlook this and proceed with Berger's account of modernization and modernity as if the modernized world, though it may be shot through with alienation, anomie, egoism, and errors, works much as it's supposed to work. If imperfections of the sort we've discussed are not unduly crippling, we may approximate the modernized world as Berger conceives it.

Clearly, political bureaucracies are designed to operate as agencies of "explicit abstraction." Fairness dictates that this is the case. To avoid particularistic treatment of individuals or groups, to the advantage of some and the disadvantage of others, participants may be effectively reduced to the status of numbers.

As mentioned in the foregoing, Berger's use of "explicit abstraction" with regard to the functioning of bureaucratic organizations is more accessible and descriptively adequate than the roughly comparable locution, "abstract logic of production," as applied to technology-intensive organizations of production. This may seem fundamentally inconsequential and not worthy of mention, but with material as complex as that being discussed, clarity and descriptive appeal are especially important.

Passivity and pathology

Returning to our discussion of bureaucratic organization, the value of any assigned number or other designation is irrelevant unless it permits identification of participants for standardized bureaucratic processing. Insofar as this set of circumstances betokens passivity and a sense of impotence, for bureaucracies these are not dysfunctional traits. For clients, however, these same characteristics may be extremely uncomfortable, with the degree of discomfort depending on the thoroughness with which clients have become accustomed to participating in an accepting fashion in a bureaucratized world (Berger, Berger & Kellner, 1973).

Presumably, all participants in a bureaucratic organization, whatever its purpose, know ahead of time that their peculiar value as an individual will be of no interest. Berger acknowledges that some may find this offensive, especially those imbued with a strong sense of themselves as multifaceted individuals who merit being acknowledged as concretely real (Berger, Berger & Kellner, 1973). Rugged individualists, perhaps.

For better or worse, taking offense at being treated in the same way as every other organizational participant may represent a failure of socialization for anyone living in a modernized social system. In any case, getting one's due from a political bureaucracy requires participation as an abstract entity, one who expects no special treatment for the specific characteristics that one happens to embody. Bureaucratic justice, in short, presumes depersonalization, a condition that, in malignant form is referred to in the *Diagnostic and Statistical Manual of Mental Disorders*: DSM-5 as depersonalization-derealization disorder (American Psychiatric Association, 2013).

It's useful to acknowledge that psychometric tools are widely used to measure depersonalization as a manifestation of psychological pathology. The intersection of bureaucratic justice with mental illness is telling. A modernized world, it would seem, requires a massive proliferation of organizations that generate sometimes debilitating disease.

Furthermore, as with technology-intensive organizations of production, bureaucracies demand repression in the form of stringent emotional control. This is yet another reason why participation in a bureaucracy may cause substantial stress. However, nearly all participants share the taken-for-granted knowledge that, much as in the workplace, open and disruptive expression of their emotional state is not suitable in a bureaucratic organization (Berger, Berger & Kellner, 1973).

The style of consciousness fostered by routine participation in bureaucratic organizations is not difficult to discern. Participants are expected to be rational, orderly, anonymous, under strict emotional control, tolerant of passivity and lack of personal efficacy, and to have an expectation of predictability. For the most part, these characteristics are consistent with the style of consciousness needed by participants in technology-intensive organizations. They also seem consistent with a prescription for tolerance of totalitarian rule, or for life in a concentration camp. Furthermore, they read like instructions for development of neurotic hysterical neurosis and other forms of repressive nervous disorder, including depersonalization-derealization disorder (Ropper & Samuels, 2009).

Compatible styles of consciousness

While the style of consciousness peculiar to a technology-intensive organization is not identical to that fostered by a bureaucratic organization, there are no characteristics with regard to which they clash. With minimal retraining, those working in one kind of organization could easily come to work or otherwise participate effectively in the other.

That this style of consciousness is pervasive in the world at large is manifest in the fact that we know how to participate in bureaucracies and we know what to expect without being taught. Beyond that, Berger shows little interest in demonstrating spillover from bureaucratic environments to the world of everyday life, seemingly taking it for granted that the spillover is abundantly obvious and quite similar to spillover from technology-intensive organizations (Berger, Berger & Kellner, 1973).

Spillover from political bureaucracies, without a great deal of further elaboration, can be construed as part of our common stock of knowledge. It's something shared by almost all of us, and is distributed more or less evenly. We take it for granted and departures from what we assume to be true would be unsettling and require a persuasive explanation.

As we have noted, however, we should acknowledge the potential for pathology intrinsic to bureaucracies and its consistency with that which we found in technology-intensive organizations. The fundamental impersonality inherent in both, as well as the basic lack of perceived personal efficacy in determining what the organizations do, may give rise to alienation from the everyday world (Marx, 1844/1988).

So long as this remains a painful characteristic limited to discrete individuals it is of no organizational concern. However, should it give rise to a hostile collectivity, the public relations or propaganda-generating capabilities of the organizations will be brought to the fore. If they fail, organizations may be in trouble, forced to make reforms deemed inconsistent with their mission (Berger & Luckmann, 1966).

We have already acknowledged that the characteristics of bureaucratic organizations have long been well known. To my knowledge, however, there has been no epidemiological research on adverse consequences, especially in the form of mental disorders. An ambitious and clever epidemiologist might very well make a prominent name for himself if such research were done, and my speculative diagnoses turned out to have merit. "Bureaucracy Drives You Nuts!" seems a fitting tabloid headline.

However, even if participation in bureaucratic organizations were found to have seriously adverse consequences, it is unlikely that much of anything would change. Berger acknowledges the likelihood of alienation and anomie as consequences of technology-intensive labor and bureaucratic participation, but as far as I can tell he accepts them as inevitable. After all, technology-intensive production and bureaucratic organizations are, as Berger understands things, primary constituents of modernization and modernity, and he expects modernization to be a painful process. Kafka's short story *Before the Law*, first published in 1915, presents this pain brilliantly. Lack of efficacy in the face of impersonal megastructures is, indeed, maddening (Glazer [Ed.], 1992).

It remains true, however, that in spite of the overriding commonalties between the two types of organizations – technology-intensive and bureaucratic – they are still, in my view but not Berger's, minimally conspicuous manifestations of pluralism. As such, though they are not implicated in the sacred/secular dichotomy, they may have marginally damaging implications for the cultural integrity of the social system in which we find them. They introduce an element of pluralism or, if you prefer, diversity, into whatever social system they are located, but their influence in this respect is limited. Nevertheless, the two types, along with variation that occurs within them, are consequential manifestations of the division of labor in society (Durkheim, 1897/1997).

12 Pluralism

In a modernized world, mass media, urbanization, and the ongoing process of technological innovation assure us that an enormous level of social and cultural complexity and an endless proliferation of choices will fractionate what once may have been a homogeneous population. A multitude of contacts with the outside world will introduce still more diversity, and ideas that were once quite literally unthinkable will become commonplace sources of tension and discord. Berger's interest in pluralism and his insistence on its importance are emphatically tied to his interests as a sociologist of religion. An institutional area, the social and cultural arena of spiritualty, may once have provided the same experiences for all, but now there are alternatives, and they be quite numerous, differing sharply one from another. As a result, an institutional realm that once was an unquestioned – even unquestionable – source of certainty is diminished as a basic foundation for truth and the comfort that may be provided by unassailable belief.

Berger has taken the position that pluralism is a precursor of secularism. Thus, if we want to understand why some modernized societies are more secular than others, we need a well-developed theory of pluralism.

In an early reference to pluralism Berger defined pluralistic societies as those that have a "shared core universe, taken for granted as such and different partial universes coexisting in a state of mutual accommodation" (Berger & Luckmann, 1966: 135). Furthermore, according to this early reference, pluralism presupposes a complex division of labor, a highly differentiated system of social stratification, and substantial economic surplus (Berger & Luckmann, 1966: 135). This suggests that pluralism means social diversity, in whatever form.

Whether or not this early rendering is consistent with Berger's subsequent call for a theory of pluralism to explain, among other things, the diffusion of secularism, remains to be seen. At first blush, however, he seems to have changed his view substantially. The primary difference between then and now is that the early definition seems clearly to apply only to modernized societies, while in his more recent work pluralism as a precursor of secularism applies to Third World countries, as well.

However, Berger's early discussion of pluralism also includes the observation that pluralism promotes and accelerates social change. This is because pluralism may undermine traditional patterns by offering readily available alternatives (Berger & Luckmann, 1966; Bickel & Dufrene, 2001). In this regard, Berger's account of the nature and effects of pluralism is much the same as that which appears in his later work and clearly helps to explain why he regards pluralism as a precursor to secularism.

Manifestations of pluralism

Thus we can also see that by juxtaposing organizations devoted to technology-intensive production with political bureaucracies we are effectively illustrating the fact that participants in modernized societies are involved in a plurality of different life worlds, though Berger seems uninterested in this dull and shallow contrast, especially since it lacks obvious religious import (Berger, Berger & Kellner, 1973). Moreover, organizations of technology-intensive production and political bureaucracies show us but two such contrasts. Taken together they are but two domains in a highly differentiated and economically substantial social system, the kind that Berger held were consistent with pluralism (Berger & Luckmann, 1966).

Nevertheless, insofar as the styles of consciousness fostered by these different social settings, while similar are not identical, we begin to see the possibility for struggle that participants engage in as they endeavor to acquire a coherent and broadly applicable way of thinking about their circumstances and their very identities. This is a problem, first and foremost, in the sociology of knowledge: an effort to investigate the concrete relationships between thought and its social context.

Lest we forget, moreover, we have completed our discussions of technology-intensive organizations of production and political bureaucracies with only occasional acknowledgement that there is a private world, the world of home and family and similarly domestic-sounding phenomena, where we may spend much of our time. As we have noted, the private sphere is influenced in a variety of ways, especially with regard to style of consciousness, by spillover from the world of work and from ubiquitous bureaucratic entities. Nevertheless, the existence of a separate and distinct private arena of consciousness and activity is another reason why it seems essential that we acknowledge that the modernized world is constituted of a multiplicity, diversity, or plurality of life worlds.

Considerations such as the foregoing may be the kind that have led Berger to the conclusion that "Instead of a theory of secularization, what is needed is a theory of pluralism" (Berger, 2011: 139), as we have acknowledged. Secularization theory is a conceptually simple perspective (Berger [Ed.], 2009). It is premised on the assumption that as societies become more modernized and rational, they become less religious, less sacred, and more secular. According to Berger, moreover, a theory of pluralization would explain why some parts of the world are more secularized than others. Berger is especially interested in his judgment that Western Europe is more secularized than the United States. It depends in part on how secularism is defined and measured, but according to most accounts Western Europe is the more secularized of the two areas (cf. Agrama, 2012).

It is striking, however, to note that we have introduced three different areas of activity – the technology-intensive organization of production, the political bureaucracy, and the private sphere – and found only modest differences in styles of consciousness among them. In good part I think this reflects the fact that they are complementary parts of the same social system, with technology-intensive production being of directive importance.

Another way of instructively saying much the same thing is that each of the three domains, as presented here, is part of a thoroughly modernized society. As such, they are mutually implicative and subject to overflow from one domain to another.

We should remember, however, that there is no all-inclusive and compelling reason why the institutions that constitute a social system need be logically consistent, a point we made earlier. Insofar as the institutions we have introduced are logically compatible, we have been able to offer only very tentative, speculative explanations as to why

Unitary traditional societies and modernized societies

Following Tönnies's (1887/2001) distinction between Gemeinschaft and Gesellschaft and Durkheim's (1897/1997) explanation of the sharp break between mechanical and organic solidarity, we have acknowledged that in traditional societies a multiplicity of life worlds was not known. A shared, undifferentiated life world was something that people took for granted. Sources of pluralism, should they be found, intruded from the outside, much as is the case with Third World countries today.

However, according to Berger, modernity means life with large numbers of strangers (Berger, 1986: 32). In modernized societies, a unitary life world provided with social solidarity due to something like Durkheim's (1897/1997) collective consciousness, has been replaced by a diversity of domains, each with its own set of rules for getting by and making sense of participants' experiences (Wiebe, 1966). This process of change raises new and troubling questions as to what ordinary people have reason to accept as real. We have seen this at least suggestively in our discussion of technology-intensive productive organizations, political bureaucracies, and the private domain.

Differences among these areas of activity, however, while real and consequential, are modest and compatible. They tend to be mutually consonant rather than discordant. This raises a question of first importance: If pluralism is as socially devastating as Berger claims, where does it come from?

From the unitary to the plural

Berger attributes the pluralization of both the public and private spheres of our world fundamentally to urbanization and mass communications. This places him at odds with Durkheim (1897/1997), the most conspicuous and influential classical sociologist to focus on the massive impact of social differentiation and the demise of a collective consciousness. As we know, Durkheim (1897/1997) attributed these processes to an ever more varied and increasingly complex division of labor, something that was facilitated by urbanization but not caused by that phenomenon.

Nevertheless, the differences between Durkheim (1897/1997) and Berger (Berger, Berger & Kellner, 1973) are not as consequential as a first look might suggest. After all, for Berger urbanization exists not only, or even primarily, as growth in geographical scope and population size. Instead, as his interest in pluralization suggests, urbanization for Berger is marked most conspicuously by dramatically exaggerated complexity, and much the same was true for Durkheim.

This applies not only to the division of labor, but to religion, ethnicity, political philosophy, and a variety of other distinctions among inhabitants of urbanized places (Berger [Ed.], 2009). An urban area, following Berger, is a "meeting place of discrepant worlds" (Berger, Berger & Kellner, 1973: 66). This is an excellent way to call forth the fundamental meaning of pluralism.

As noted with reference to the disintegration of Durkheim's collective consciousness, an urbanized world is characterized by multifaceted diversity in cognitive content and styles of consciousness (Durkheim, 1897/1997). Moreover, the influence and encroachment of an urban way of life and style of consciousness need not be limited to urban areas. Mass media make it ubiquitous.

Mass media and an urbanized world

The proliferation of media of mass communications assures that even those living in rural areas, perhaps on a working family farm, will become imbued with an urban approach to living and making sense of their circumstances and the rest of the world. The culture of the city – an urbanized culture – thereby comes to characterize an entire society, including non-urban areas and their inhabitants (Berger, Berger & Kellner, 1973), though my experience suggests that Berger exaggerates this process and the urbanization of styles of consciousness of geographically rural people (Bickel & Howley, 2003).

While I am in basic agreement with Berger's position, I have done substantial research and program evaluation in both rural and urban areas. Berger comes dangerously close to asserting that differences in ways of living and style of consciousness have entirely disappeared between these two contexts. While there are more similarities than differences, the two settings are far from being identical. The characteristic that struck me most forcefully was rural residents' determination to maintain a rural residence, one that keeps them close to and in continuing interaction with their families of origin (Bickel, Smith & Eagle, 2002).

Nevertheless, as many others have noted, the growth and influence of mass communications assures that everyone located in a particular social system is constantly exposed to an almost maddening diversity of often conflicting information. Under such circumstances, it is difficult to imagine anything sacred that is not, even if only indirectly and without anyone's intention, subjected to contradiction. Instead, there exists an overwhelming plurality of perspectives, points of view, and styles of consciousness.

If we take Berger's account seriously, it becomes easy to see why he is convinced that pluralism precedes and accounts for secularism. The sacred survives as such only if it is manifest in the lives of members of an intact group. In an interesting and remarkably useful turn of phrase, Berger terms these groups plausibility structures, collectivities for which the sacred is real and by which it is sustained. Under conditions such as those just described, namely conditions suffused with a plurality of images and accounts, the integrity of such supportive groups is very likely to be lost. Again, it seems reasonable to judge that the sacred becomes secularized because of the ascendance of pluralism.

The unfolding of these processes in any society raises problems of legitimation. We can certainly see this in the contemporary United States where differences in race, ethnicity, religion, class, and various other factors threaten to terminally fracture the existence of one society and foster nativism and xenophobia (cf. Habermas, 1975).

When prominent and powerful representatives of member states threaten, even if disingenuously, to adopt the mid-nineteenth-century strategy of seceding from a union that, they claim, has become alien to their way life and cultural heritage, the need for more effective legitimation is made clear. Calls to celebrate diversity, or patriotic claims that our democracy has always been characterized by diversity, seem likely to be ineffective (Bickel, 1997).

One way of legitimating a social system, notably a system that has become extremely diverse and includes what appear to be educational and other avenues of upward mobility, is to appeal to a real or imagined array of desirable opportunities open to almost everyone. *The American Occupational Structure* (Blau & Duncan, 1967) is an early and persuasive affirmation of the ascendance of educational attainment over family social class in the status attainment process. This is the sort of richly propagandized social message that prompts parents to tell their children, without awareness of an enormous degree of exaggeration, that they can be anything they want to be. Even if this sort of well-meaning encouragement were highly likely to be true, it would be yet another source of diversity in an already diverse society.

Similarly, Berger makes clear that the forces of legitimation and division that are linked to social class emphatically include religion. Class location is a powerful predictor of religious affiliation, and simultaneously of factors usually closely tied to religion such as views with regard to abortion, homosexuality, capital punishment, birth control, same-sex marriage, and numerous other more and less controversial and timely issues (Berger, 1992). The complexity of shifting relationships among class, religion, and a collection of specific, politically charged contemporary issues has an increased density in a modern world characterized by a diversity of means invoked in an effort to gain legitimacy.

In Berger's view, for example, if the Christian church, seeking legitimation, gives up its central tenet of redemption through Christ to more easily accommodate the taken-for-granted wisdom of the everyday world, its legitimacy is lost rather than maintained (Berger, 1992). Berger is cautious, however, to avoid adopting an unyielding orthodoxy. Berger's position is clear, but so is his aversion to the sort of fundamentalism that, in an effort to restore lost taken-for- grantedness, prejudges all issues, leaving nothing for civil discourse (Berger, 2014). Berger explains his point of view as follows:

> The very last thing I want to advocate is some sort of fundamentalism or even orthodoxy. I am suggesting that there is a middle ground between the "progressive" surrender to the age and the fundamentalist and denial of it (Berger, 1992: 16).

The notion of a middle ground to which human beings unself-consciously gravitate when dealing with contentious issues is, as we have already suggested, something that Berger takes very seriously. It enables us to come to terms with the differences that separate us and do so without sacrificing that which we hold sacred. Moreover, by its very nature it fosters a mature and uncorrupted understanding that truth is never perniciously absolute or narrowly exclusive to the detriment of decency and goodness and the rejection of overarching contextual factors.

Contextual factors, while a forceful presence, as in the form of movements toward modernization, must be examined critically rather than weakly accepted as inescapable (Berger, 1992). Thus, an environment characterized by diversity, with a multiplicity of sometimes vastly different choices, repeatedly calls forth our most effective critical faculties. Our examination of social circumstances is instructed by acceptance of the socially constructed nature of what we take to be real (Berger & Luckmann, 1966).

At the same time, however, we must recognize that the wholesome middle ground must be sufficiently well-defined and fraught with untroubled conviction that it provides us with a sense of stability. This stands in sharp contrast with the thoroughgoing relativism that has been adopted by many and may be construed as a source of freedom, unfettered by dogma and what we may take to be wholly arbitrary and stifling moral codes. However, freedom can take an excessive, discomfiting form, and we may find ourselves in an anomic and egoistic set of circumstances, with nothing to believe in, nowhere that we belong, and no reason for doing one thing rather than another (Durkheim, 1897/1997 & 1912/1995).

When certainty is hard to find in a domain marked by diversity, one that offers too many choices with too little guidance, we may find ourselves turning, perhaps for the first time, to religion. The religion that we adhere to may be one that demands substantial discipline in an effort to secure the confidence, conviction, and trust without which we may find only free-floating misery. For the most part, however, following Berger, if we turn to religion we will tend toward a middle ground.

Technology-intensive productivity and upward social mobility

Berger takes the position that wherever increases in technology-intensive productivity occur, the incidence of social mobility increases, most of it upward (Berger, 1963: 49). Nevertheless, I find it interesting and to Berger's credit that he acknowledges that the degree of upward social mobility is commonly exaggerated, certainly in the United States, where many still hold the patently erroneous view that upward social mobility in America exceeds that in all other modernized countries (Putnam, 2015).

The mistaken belief in massive and consequential levels of upward mobility contributes to forming the character of participants' style of consciousness, and it has a powerful bearing on patterns of anticipatory socialization (Berger & Huntington, 2002). Whether or not the clash of expectations with reality turns out to infuse participants with a keen sense of disappointment, loss, and even alienation is an issue that Berger does not address.

It seems likely, however, that when people who believe this sort of overblown cheerleading see that their achievements don't measure up, their attendant feelings of failure are apt to foster sadness, vulnerability, feelings of inferiority, and perhaps a willingness to rebel through recourse to right-wing populism. This may be accompanied by the forlorn conclusion that they have shortchanged those closest to them (Newman, 1993 & 1999).

When feelings of this potentially despondent kind are joined with the alienation, anomie, and egoism that come from routine participation in the primary institutions of modernized societies, serious social problems – if not outright rebellion – seem sure to eventually emerge. It's sometimes difficult to understand how the continuing confluence of disappointments and taken-for-granted depredations has not given rise to even greater social difficulties. Perhaps a protective posture with regard to the ubiquitous nuclear family has been a force for caution and restraint.

By way of making clear that when a participant's nuclear family is taken into consideration things become more complex and potentially unsettling, Berger tosses in one of his clumsier neologisms: "multi-relational synchronization" (Berger, Berger & Kellner, 1973). As far as I can tell, this just means that if the futures of two spouses and their children are simultaneously acknowledged, decisions become even tougher. No doubt this is true, and it's something that merits being explicitly stated. But this is easy to do and its implications are easy to discern using conventional language. What is gained by inventing "multi-relational synchronization" is an etymological mystery.

The encroachment of pluralism

It's certainly true, nevertheless, that in a less diverse society this sort of complication would be unlikely to emerge. Among other things, the place where one was born might be invested with a salutary significance, much as I observed in rural Appalachia (Bickel, Smith, & Eagle, 2002). For better or worse, this kind of influence tends to hold one in a familiar location, rather than occasioning dispersion across the continent, or perhaps throughout the world.

Furthermore, in the absence of wholesale diversity, planning for the future could be greatly simplified. If a husband plans to stay put and do the same work his relatives have done, while his wife stays home and keeps house and raises the children, there is likely to be a lot less to plan for. Life is still unpredictable and precarious, but more things stay the same. The encroachment of pluralism is significantly forestalled and its consequences are largely deflected (Newman, 1993).

Compare the two sets of circumstances, the highly pluralized and the traditionally simple or unitary. It seems clear that, as presented by Berger, the pluralized condition will be favorable to secularism – there are just too many choices, too much cross-group communication, and too many complications for it to be otherwise, or so it seems. The simpler set of circumstances, on the other hand, will be a good deal more hospitable to the traditionally sacred, providing a much stronger sense of normative stability and belonging. In Berger's view, the highly pluralized context promotes individual freedom, but it comes with costs, including the by now-familiar social conditions of anomie, egoism, and alienation (Ollman, 1977).

It is easy to see why Berger has concluded that the encroachment of pluralism is significantly forestalled and its consequences are largely deflected (Newman, 1993).

13 Pluralism and capitalism

It is at this point that Berger's affinity for capitalism and the source of his strong preference for that economic system over socialist or other alternatives becomes startlingly clear. Though he doesn't put it in just these words, Berger's travels through capitalist countries in East Asia convinced him of the superiority of capitalism in generating modernization in the form of economic growth and widely shared prosperity (Berger, 2002 & 2011).

Though Berger eschews Utopianism, his uncritical commitment to capitalism brings to mind one who is thoroughly imbued with a utopian vision tied to a specific mode of economic organization (Berger & Kellner, 1981; Berger & Huntington, 2002; Berger, 1986, 1990 & 2011). This judgment is magnified by Berger's failure to distinguish one capitalist venture from another. For him, they are all presented as roughly the same, and they are all presented as better than any alternative.

Defining capitalism

Berger is not one to think in terms of capitalism as a system whereby capital exploits dependent labor to the advantage of individual capitalists and capitalist organizations (Marx & Engels, 1844/1988). In addition to being ideologically at odds with Berger's views, it seems certain that he would object to the word "exploits" as being vague, unmeasurable, and value-laden, not suitable for sociological discourse. Berger suggests very brief but useful definitions for capitalism, such as an economic system where exchange is regulated by market mechanisms. However, as we have seen, he ultimately settles on a simple but astonishingly hopeful and almost recklessly wholesome alternative, specifically "production for a market by enterprising individuals or combines with the purpose of making a profit" (Berger, 1986: 19).

One might readily observe that the word "enterprising" gives rise to exactly the same objections as the word "exploits," only more starkly. Berger, nevertheless, after a good deal of deliberation, includes it in his definition of capitalism.

Marx, however, devised a measure he called the "rate of exploitation" (Bickel, 2013). Anyone who has had reason to use the rate of exploitation, even as a self-instructional exercise, will immediately recognize that Marx purported to be able to measure exploitation in a simple, straightforward, and seemingly precise way. The measure itself, moreover, has a good deal of intuitive appeal. One may take issue with Marx's offering, but his use of exploitation gives substantial evidence of much more sustained and disciplined scientific effort than Berger's unlikely and obviously unsuitable use of "enterprising." I suppose it would be possible to devise a measure called the "rate of enterprise," but to my knowledge no one has, nor does such a development seem on the verge of happening.

Nevertheless, gone from Berger's discussion of capitalism are ideas central to a Marxist conception such as class antagonism, the natural price of labor, commodity fetishism, socially necessary labor, and many more (Marx, 1867/2011). Berger's sanitized rendering of capitalism may be superficially plausible but it's also dangerously simplified in a way that would make the most ruthlessly avaricious capitalist blush at its bland, non-conflictual rendering. Berger's cloying use of the modifier "enterprising" reduces his definition to the status of the simple minded, though Berger is hardly simple minded.

Much earlier in this manuscript we made the seemingly safe observation that Berger rarely presented ideas in a vacuous and simple-minded way. Berger's stunningly sophomoric take on capitalism may be one of those rare occurrences.

Elsewhere, however, Berger includes fairly long and useful lists of characteristics of capitalism that are much more adequate than his very brief and uninformative definition. His distinction between economic systems driven by market mechanisms as opposed to planned allocation is decisive in making the nature of capitalism clearer by comparing its economic decision-making process with an essential aspect of socialism. In just a few worlds, he surpassed the explanatory power of his cheerleading definition based on "enterprising" individuals alone or in combination (Berger, 1986).

Berger also recognizes that a capitalist world system, in nascent state, existed in the nineteenth century and has continued to evolve at an accelerating rate in the twentieth and twenty-first centuries. He doesn't develop this idea to the extent that recent research and theorizing warrants, but in contrast to many others writing in the latter half of the twentieth century, he clearly acknowledges the existence of a capitalist world system, as well as its specific cultural character (Berger & Huntington, 2002: 1–16).

Capitalism and socialism

In addition, Berger is openly forthcoming with regard to a fundamental difference between capitalism and socialism. While capitalists produce for profit, socialists produce for human use, bringing to mind the distinction between exchange value and use value frequently invoked by Marx (1844/1988), and that we earlier purported to find manifest in Weber's concepts of Zweckrationalilty and Wertrationality (Roth & Wittich, 1978).

To many, this distinction may seem to give the greater moral authority and material legitimacy to socialism. Whether or not this is the case, Berger's willingness to express the difference between these two different systems for harnessing and directing technology-intensive production in the way he did is a laudable mark of scholarly even-handedness. Berger could have easily invoked the distinction between market mechanisms and mechanisms of political allocation without intimating, at least in the view of some readers, that socialism may have a more humane purpose than capitalism. But to his credit he chose not to hedge by playing it safe.

Economic culture

Berger subsumes his own theory of capitalism under the heading "economic culture" (Berger, 1986: 7), a term we introduced earlier to refer to the non-economic implications and manifestations of capitalism. According to Berger, such a theory would "explore the social, political, and cultural matrix or context within which [the] economic processes [of] capitalism operate" (Berger, 1986: 7). In other words, Berger is interested in identifying contextual factors essential for capitalism to prevail and that capitalism inevitably brings with it. This is what we would expect of a good sociologist with interests suitably intrinsic to his discipline.

Berger also uses the idea of economic culture to identify corruptions of ostensibly capitalist systems that render them something else, say a last vestige of feudalism in the late twentieth century. This kind of reasoning leads Berger to reject the claim that South Africa was a genuinely capitalist society before the elimination of apartheid. Thus, while no economic apparatus is entirely consistent with one mode of production (Poulantzas, 1978/2015), the apartheid-driven feudalism of South Africa was so conspicuous and thoroughly inscribed in its society and its economic life that the designation capitalist did not legitimately apply. Berger prefers the characterization "racial feudalism," with no recognition of capitalist components (Berger, 1986 & 2011).

If apartheid had been only loosely tied to the South African economy, and apartheid had been only incidental to production and to the composition of the labor force, the term capitalism would have been much more fitting. As we know, however, apartheid was an integral and essential part of economic organization as construed and controlled by the ruling white South Africans.

At first glance, I had thought that the economic culture terminology vindicated my inclination to characterize Berger as a cultural determinist. This is certainly consistent with the cultural emphasis in Berger's introduction to *Many Globalizations* (Berger & Huntington, 2002). However, as it became clearer just what Berger wanted economic capitalism to do – address capitalism and social stratification, capitalism and forms of political organization, capitalism and various social and political value systems, capitalism and cultural patterns, capitalism and cognitive structures – I realized that I was mistaken (Berger, 1986).

The meaning of economic culture expands, including more substance, as Berger works through *The Capitalist Revolution* (Berger, 1986). Throughout, however, he is concerned primarily with the non-economic manifestations of capitalism, and the kind of society they produce (Berger, 1986).

Fifty hypotheses

Of central importance in *The Capitalist Revolution*, Berger sets for himself a difficult task: formulating and, in a preliminary manner, evaluating fifty hypotheses as part of an ongoing investigation of the nature and concomitants of capitalism (Berger, 1986). In some respects, it's like nothing I've seen before: detailed, ambitious, uncertain, very poorly executed, and covering a fairly broad swath of history. Fifty propositions is a lot, especially when they are to be empirically evaluated "in the strict sense in which this term has been use in the social sciences" (Berger, 1986: 3). In truth, however, there turns out to be nothing "strict" or rigorously formulated about the loosely structured evaluation that Berger produces with his fifty propositions.

In Berger's judgment, it seems overwhelmingly evident that capitalism's payoffs in material abundance give rise to the most comfortably habitable modernized societies and the most rewarding lives (Berger, 1986). Whatever the social and cultural differences among societies where capitalism takes hold, and the differences are numerous and consequential, prosperity overrides all other considerations as modernization proceeds, or so Berger has concluded in an unspoken but unmistakable way. In fact, as we have already intimated, for Berger prosperity and the benefits of modernization are, in large part, coterminous (Berger, Berger & Kellner, 1973; Berger & Huntington, 2002). Perhaps that's just good common sense, but it strikes me as a gross oversimplification.

None of us is obliged to share this view. After all, it's not intrinsic to the definition of modernization that Berger gave early on: "the institutional concomitants of technologically induced economic growth" (Berger, Berger & Kellner, 1973). At this stage in the development of his thinking, however, he apprises us of the heart of modernization as he sees it, and it is clearly a product of mature capitalism and its technological-industrial attainments.

In fairness to Berger, it is important to emphasize that Marx himself had genuine admiration for the innovative nature and productive power of industrial capitalism (Marx, 1867/2011). As we know, in Marx's view, a well-developed capitalist economy was a prerequisite for the establishment of socialism (1844/1988). In that sense, capitalism has been a revolutionary force for social change in more ways than one (Schumpeter, 1942/2008).

Whether for the long term this is for better or worse is subject to contestation. It is a difficult issue to address, because "in the long term" is inherently vague. There is no doubt, however, that capitalism's revolutionary character as a force for economic development has been quite real (Berger, 1986; Berger & Huntington, 2002). That there is more to capitalism, and that much of it is quite destructive, is something Berger acknowledges and that we will cover as we proceed (Berger, Berger & Kellner, 1973).

Berger's propositions and their substantive and methodological merit

This brings us to Berger's first two propositions. We wish to make clear at the outset that though large parts of *The Capitalist Revolution* consist of an anti-Marxist polemic, it is, as noted, primarily a book about the economic culture of capitalism. Nevertheless, we will not discuss all fifty propositions, limiting ourselves to just a few that do a creditable job of explaining Berger's sociological understanding of capitalism. On that basis, I think the first two propositions are eminently well-suited to beginning our discussion.

Since Berger does not number his propositions, a seemingly sloppy and obvious oversight that inconveniences the reader, we will number them in the order they occur in *The Capitalist Revolution* (Berger, 1986: 36):

> P1: Industrial capitalism has generated the greatest productive power in human history.
>
> P2: To date, no other socioeconomic system has been able to generate comparable productive power.

First things first: Are the propositions testable? Since Berger offers nothing specific to which the propositions would be compared, it's clear that they are not. "Greatest" and "no other" are strong locutions, but without a specific basis of comparison we're left asking "With respect to what?" And in that regard Berger is not forthcoming.

As far as I can tell, moreover, these two propositions mean the same thing. Why Berger included both is a mystery. Perhaps it was for emphasis, though that hardly seems necessary and the presence of two propositions that mean the same thing generates needless confusion. In any case, whatever our response to one proposition, our response to the other will be the same.

Both, moreover, are consistent with Berger's claim that capitalism has constructed the material circumstances that have yielded the full-scale modernization of a large part of the world (Berger, 1986). So, while we can see pretty well what Berger is getting at, his failure to put the propositions in proper form, subject to interpretable comparisons, has rendered them untestable, even though rigorously testable propositions are the kind he promised.

This is a set of circumstances that we will encounter with all of the propositions we address, as well as those throughout the rest of the book. We have no trouble understanding what they mean, but they are not presented in testable form. They may be excellent for the ongoing work of a sociological tourist, bearing witness to a good deal of informed thought, but they lack the form and reasonable level of precision needed to be testable.

Clearly, both statements apply to the United States and Western Europe, as well as the Asian Crescent of Prosperity, but I think the jury is out for most of the rest of the planet, though Berger would take a more expansive view including, for example, Jamaica (Berger, 1986 & 2011). In any event, given Berger's understanding of modernization, capitalism seems clearly to be unparalleled in its promotion of modernization and modernity in the Western World and, more recently, in East Asia (Berger [Ed.], 1990; Berger & Huntington, 2002).

One can agree that both propositions have been historically accurate for the West and other areas we have just identified, and still not be in agreement with Berger's understanding of the best way to promote modernization. After all, we don't have to share the view that prosperity and modernization are coterminous. Some of us may give priority to moral values, aesthetic tastes, or spiritual attainments as manifestations of modernization (Berger & Huntington, 2002). Berger understands this, and his account of his conversion to the point of view of a thoroughgoing capitalist, while for him no doubt momentous, is not as convincing in its observational details as one might expect.

I don't doubt the conversion, just the persuasiveness of its genesis. One can only wonder at the lack of consistency with the multiple-vantage-point debunking motif that was so irrepressibly conspicuous in his early work and ask, "is capitalism really that effective for all times and places?"

Yes, we've agreed that, at least in the West and East Asia, in the aggregate it has been. But if we want to tie this to modernization, it is a rare modernizing project, indeed, that exists on a world scale. Small scale projects are commonplace, operating under a variety of auspices (Hickling-Hudson & Klees, 2012), and they can be capitalistic or socialistic in nature, or the work of those whom we have referred to as non-denominational modernizers (Berger, Berger & Kellner, 1973). (The concept denomination here is used in an informal, commonsensical way, and should not be confused with Berger's use of denomination as a form of religious organization [Berger, 2014].) Berger understands the complex nature of modernization quite well, and one wonders if he has become a very different sociologist and person than the one who in *Invitation to Sociology* based sociology's value primarily on scientific integrity (Berger, 1963).

Beyond that, if we are to make decisions as to the comparative efficacy of capitalistic, socialistic, or other modernization endeavors, it is essential that we acknowledge that placing them under the same label does not mean that each does not exhibit substantial variability. Perhaps we are straying a bit too far from the first two propositions, but at any level of analysis we may find greater and lesser instances of productivity among capitalist efforts and greater and lesser instances of productivity among socialist efforts, and so on. Furthermore, I would be stunned to find that there was no overlap. There might even be outliers: one or more socialist ventures that surpassed in productivity each and every capitalist arrangement.

With the collapse of Soviet socialism, comparisons such as those just introduced may be of much less interest than before 1989. However, writing in

1986, Berger invites comparisons by using phrases such as "greatest productive power," which he has good reason to do. It is unclear, however, as to whether he is comparing programs initiated under the auspices of capitalist and socialist societies, or if the capitalist and socialist designations apply to the organization and functioning of the arrangements themselves. Conceivably, a socialist modernization undertaking could have been made under the nominal auspices of a capitalist country.

In an unexpected and problematic assertion, Berger attributes at least part of the productive power of capitalism to the fact that it provides a comfortable, even nurturing environment for both entrepreneurs and those with interests and skills better suited to engineering (Berger, 1986). As is too often the case, Berger doesn't develop this assertion, but for him, clearly, it is unassailable.

Surely, however, Berger is familiar with Veblen's work regarding the clash between profit-seeking entrepreneurs and technocratically favored engineers and craftsmen (Veblen, 1904/2014). Similar arguments were very effectively made by the brilliant but less well-known chemist turned historian of science, David Noble, in his books *America by Design* (Noble, 1979) and especially *Forces of Production* (Noble, 1984).

Veblen's argument, presented in widely read volumes such as *The Engineers and the Price System* (Veblen, 1921/2013), *The Instinct of Workmanship and the State of the Industrial Arts* (1918/2015), and *The Theory of Business Enterprise* (1904/2013), holds that capitalists' narrow pecuniary interests constrain the ingenuity and productive contribution of those who work as engineers, technicians, and in similar occupations devoted to developing innovations that will improve the material prospects of the population at large. Berger may be in complete disagreement with Veblen, but given the latter's influence and broad readership, Veblen is not someone whom Berger, as a thorough and remarkably well-informed scholar, can reasonably ignore, a point we made earlier. I have long revered Veblen, and I would like to see Berger present his counter arguments with regard to the relationship between Big Capital and practical, workmanlike endeavors (Veblen, 1923/1997).

Berger, however, does not address that issue, making this a good place to return to more directly addressing his propositions. As already emphasized, Berger claims that his fifty hypotheses are testable. Given the for-all-time-and-all-places nature of his first two, however, that claim is problematic. Hypothesis testing is first and foremost a process of making formally interpretable comparisons. Since that is the case, Berger should provide a basis for comparison regarding his propositions as to the efficacy of capitalism. This is the sort of problem that, as we shall see, renders Berger's propositions non-testable, diminishing the credibility of his book (Berger, 1986). These are not the only recurring difficulties with testing Berger's hypotheses, as suggested by P3 immediately below.

> P3. An economy oriented toward production for market exchange produces the optimal conditions for long-lasting and ever-expanding productive capacity based on modern technology.

Early on, Berger told us that modernity was an historical concept that could be studied as such, with a beginning, duration, and an end (Berger, Berger & Kellner, 1973). We can only conclude, therefore, that modernity has little or nothing to do with production for market exchange and, it seems, that such a productive apparatus is not an historical concept. On both counts, our conclusions are based on the assertion that under the conditions specified productivity is "ever-expanding." Testing hypotheses about something that never ends is clearly impossible: How can we answer questions concerning duration with regard to something that just keeps going, never to come to a halt? Perhaps we could consult astronomers studying the ever-expanding universe, but to what end.

There are other difficulties with this ostensibly testable hypothesis that are just as easy to see and just as troublesome. The term "long-lasting" is vague, so vague that in and of itself it renders the hypothesis untestable. Furthermore, as we have noted with P1 and P2, testing an hypothesis is based on the presumption that the test involves comparison with another hypothesis. As with P1 and P2, however, no alternative is specified for P3.

The implied alternative, presumably, is any other form of production in any other place, but how many other forms and venues are there, and what is their specific nature? Berger could have specified something like "oriented toward politically directed distribution," referring to a socialist economy, but he didn't bother, so we're left with the sort of ambiguity that is intolerable when constructing testable hypotheses. Notice, again, that this is not the unreasonable demand of a methodological purist, but a simple, practical requirement of conventional hypothesis testing, something that has been understood for more than a century (Blalock, 1981).

It may well be that if we discuss too many hypotheses and give even a cursory gloss to the errors Berger has made that render them untestable, the process will become annoying. But as we have already acknowledged, we plan to discuss just a few of propositions that seem pertinent to an overview of Berger's work.

In the present instance, if we look past the obvious and insuperable barriers to testability and try to discern the substantive meaning of the proposition, we will come up with something like "A market economy makes more productive use of modern technology than a command economy." Berger, of course, would agree, all the more emphatically now that he has discovered capitalism. To his credit, however, he acknowledges that even if his proposition were unassailably true, it raises empirical questions of distribution.

The American economy, for example, produces more from year to year, but for at least the past two decades, any value that has accrued from the added productive power flowed disproportionately – in fact almost exclusively – to the very wealthy, with those of lesser means gaining little or nothing (Domhoff, 2013). The process of distribution has gotten more astonishingly inequitable as the years have gone by, with so much accruing to the very wealthy that America is more and more frequently characterized as an oligarchy or a plutocracy.

This set of circumstances stands in sharp contrast to the post-World War II era, ending in the early 1970s, during which America developed a sizable and prosperous middle class, and members of those we might describe as working class were not, by conventional standards, poor. This was a time when the value that accrued from productivity gains was distributed much more equitably, and the claim that the very rich ruled the country in plutocratic fashion was rarely made by those in the mainstream. Today this judgment is commonplace (Klein, 2007).

According to Berger, the growth of a large and prosperous middle class with substantial political clout during the 1950s and 1960s was due to the simultaneous occurrence of a large number of suitably well-educated adults and a demand for their services (Berger, 1986; Bickel, 2013). For reasons that are not clear, Berger gives priority to the abundance of employable people with substantial education and experience. In other words, his tacitly framed model of the salutary development of an American middle class offers first a supply-side response to the question to the hypothetical question "how did a substantial middle class come to be?" Nevertheless, as we will acknowledge below, the fact that Berger also stipulates a role for demand in promoting social mobility puts him in good company.

However, by giving priority to the supply of well-educated aspirants, Berger offers the sort of response that one would expect from a contemporary economist who subscribed to human capital theory (Becker, 1964/1994). Following proponents of human capital theory, if human capital, especially in the form of educational attainment is abundant, those with money will invest in it, much as they would invest in capital in any other form. Thus, given an aggregate of people with relatively high levels of human capital, investment will create jobs that will employ the human capital in a way that yields attractive returns. This supply-side model is the one that Berger finds most congenial, and he presents it, without giving it any sort of designation, as a framework for processes that everyone takes for granted (Berger, 1986).

However, I think that the assumption that human capital prompts investment is at best dubious (Bowles & Gintis, 1976). A more plausible account of the development of the American middle class gives pride of place to demand for employees in a labor market where jobs that pay reasonably well are plentiful and human capital, while abundant, does not surpass the number and kind of employees that are needed (World Economic Forum, 2015).

Unfortunately, beginning in the mid-1970s technological advances, coupled with the acceleration of globalization, undercut the willingness of wealthy entrepreneurs and corporations to invest in domestic job creation. Better for them to invest in human capital that exists in abundance in large, less-developed countries such as India and Pakistan, where wages and salaries are much lower and expectations for benefit packages, insofar as they exist, are much more modest (Bowles, Edwards, & Roosevelt, 2005).

This joined with the already discussed ever more intensive concentration of wealth and income in a small stratum of people and corporations has yielded a sharp contraction in the availability of jobs with post-World War II levels of compensation. Payoffs for individual investments in education have been dramatically diminished, and job creation, all too often, is limited to minimum-wage positions that pay too little to make ends meet (LeRoy, 2005).

It may be unduly dramatic and inconsistent with correct usage of the terminology of the philosophy of science (cf. Kuhn, 1962) but I will, nevertheless, offer the judgment that a familial paradigm shift occurred during the 1970s. No longer could a single breadwinner, typically employed by large productive organizations as well as large state bureaucracies, provide a comfortable life for a family. Instead, this level of economic attainment now applied only to those firmly situated in the upper middle class (Newman, 1993).

Women increase their participation in the labor market

The deterioration of the labor market for most participants was, for a time, hidden by the sudden and celebrated increase in the number of women, including wives and mothers, who had full-time jobs outside the home. Those who interpreted this as a formidable instance of women's liberation missed the mark for all but those in professional positions who wanted to work. The others, the less advantaged, less well-educated, the more conventional worked because they had to, commonly augmenting the income of a husband and keeping a family fed, clothed, sheltered, and intact.

When Berger wrote *The Capitalist Revolution* (Berger, 1986) he should have known this. He tells us that the book was long in the making, beginning in the 1960s, but given the book's title, it's difficult to imagine that he was oblivious to nationwide transformations that received a great deal of publicity and consumed a substantial amount of scholarly ink. Berger may have wanted to believe that America was a predominately prosperous county in the fashion of the 1950s and 1960s, but by the time *The Capitalist Revolution* (Berger, 1986) was published, this already was a passing fantasy. Perhaps recognizing this transformation would have uncomfortably shaken his unbridled faith in capitalism (Berger [Ed.], 1990), After all, his fifth proposition, presented below in the section titled "Poverty is Relative: A Stupid Supposition," is an unqualified testimonial to the beneficence of capitalist largesse.

Furthermore, after acknowledging that he could no longer take an even-handed view of competing capitalist and socialist models of development, Berger attributed this to the success of capitalist developmental efforts, especially in Hong Kong, Singapore, South Korea, and Taiwan (Berger, 2011). Perhaps Berger's commitment to capitalism is an unintended and heavily ironic example of what he terms "dual citizenship." In this instance, however, he seems to have dispensed with duality, jettisoning the sociologist's relevance structure in favor of an emphatic, no sociological sacrilege barred, embrace of capitalism.

Poverty is relative: a stupid supposition

It's almost inevitable that when Berger discusses poverty in early to mid-nineteenth-century England and acknowledges the horrors documented by Engels in *The Condition of the Working Class in England* (Engels, 1845/2009), he'd counter with one reference, Gertrude Himmelfarb's well known work *The Idea of Poverty* (Himmelfarb, 1984). Scholars on the right who relish dismissing poverty as something thoroughly relative and probably born of envy, more in one's head than one's belly, and devoid of truly damaging substance have gotten a lot of mileage out of Himmelfarb. Given the real and undeniable material difficulties of life as described by Engels and others, affluent intellectuals' conceptual exercises as to what poverty is seems obscene (Gaskell, 1833/2010).

I'm certain that Berger knows that he can't dismiss universally respected and admired work such as Engels's just by tossing in a citation from Himmelfarb. But he proceeds as if he could. I think it's a cheap way to go. I often disagree with Berger, but I rarely question his integrity. In this instance, however, I am inclined to do just that. At best he's careless, sloppy, and lacking in scholarly discipline, at worst he's craven.

The favorable effects of the industrial revolution were not felt in fully developed form in England until well into the second half of the nineteenth century (Berger, 1986). Having acknowledged this Berger, in my view, has a scholarly obligation to use more than one weak, wordy, conveniently overused reference as a means of dismissing Engels's work, pertaining to the year 1844, as "highly partisan and hardly concerned with dispassionate weighing of the evidence" (Berger, 1986: 40). How could Berger know this? We are given no reason to believe that Berger has concerned himself with Engel's dispassionate weighing of the evidence or hid failure to do so (Berger, 1986). In fact, I got the impression that Berger had not read Engel's book but that he simply dismissed it out of ideological convenience and with the tacit backing of Himmelfarb (1984).

Where is the substantiation – any sort of substantiation – for Berger's claim. The world of working people in nineteenth-century England was a truly miserable place, and the monarchy was happily oblivious and really quite profligate (Hubbard, 2013). Queen Victoria, in Hubbard's long scholarly treatment, comes across as selfish, insular, vulgar, and caring for no one but her dead husband. I'm not sure she knew there was a working class, and if she did she was not inclined to think about it.

To his credit, however, Berger does conspicuously acknowledge that the benefits of the industrial revolution did not reach the lower classes during the first half of the nineteenth century, during which Engels researched and wrote his book. Thus, the following proposition:

> P4: The early period of industrial capitalism in England, and probably in other Western countries, exacted considerable human costs, if not in an

> actual decline in material living standards, then in social and cultural dislocation.

I'm not sure why Berger offered this as one of his, in principle, testable hypotheses about capitalism. He characterizes it as tentative, and repeatedly admonishes the reader that the relationship implied by the written description may be far from accurate, possibly taking a wide variety of forms (Berger, 1986: 41 & 42). Furthermore, he warns us that the relationship may bear little or no resemblance to that which applied to other countries at other times.

In addition, by using the imprecise words "early" and "probably" as in "The early period of industrial capitalism in England and probably in other ...," Berger has rendered the hypothesis untestable. Unless there is precise, beforehand agreement on the value of "early," and we get rid of the imprecise word "probably," that is, indeed, the unhappy set of circumstances in which we find ourselves, and the vagueness of the word "considerable" just makes things worse.

This is also an unsuitable two-for-one proposition. When Berger argues that early industrial capitalism "exacted considerable human costs, if not in actual decline in material living standards, then in social dislocation," we are presented with two hypotheses, and we don't know which one to test: the one about "human costs" or the one about "social dislocation," or both.

Nevertheless, I take this material as a welcome effort to be accurate and evenhanded with regard to the costs of capitalism. As it's written the two-for-one hypothesis is not testable, but with a little work, it could be made so, in the form of two testable hypotheses. For one with an interest in how capitalist economies and the societies that constitute their developed forms emerge, this is an interesting, well-placed issue, and Berger is acknowledging that capitalism has been beset by imperfections, something I had thought he no longer believed.

In addition, P4 provides an interesting juxtaposition with P5. The substantive comparison of the complementary propositions P4 and P5 is striking, the sort of thing that an experienced and capable writer such as Berger does:

> P5: Advanced industrial capitalism has generated and continues to generate the highest material standard of living for large masses of people in human history.

Again, however, Berger does himself in with imprecise language. Notice his usage of "highest" as in "highest material standard of living" and "large" as in "large masses of people" Thus we beg the questions "highest" with respect to what and "large" with respect to what? In addition to producing substantive confusion, the vagueness of these terms renders P5, in both parts, that which refers to "highest" and that which refers to "large," untestable. Again, this is contrary to Berger's promise that the conditional propositions or hypotheses would be empirically testable in a strict social scientific sense (Berger, 1986).

Furthermore, P5 is so similar to P1 and P2 that it introduces still more redundancy.

Berger's decades as a sociological tourist were eminently productive. But we have had reason before to surmise that they left him careless and unwary of the pitfalls that open up when one resorts to more formalized methodological procedures, even those of a relatively simple kind. Berger's work is still informative and worth reading, but it is also instructive as a repository of methodological issues.

At this point it's probably unfair to add to our account of Berger's methodological limitations by taking material from a book other than *The Capitalist Revolution* (Berger, 1986). However, I think this one error may suitably amplify the costs of spending one's professional life doing the otherwise edifying work of a sociological tourist.

In *The Many Altars of Modernity* (Berger, 2014), Berger makes the following claim:

> [P]luralism, the co-existence of different worldviews and value systems in the same society, is the major change brought about by modernity for the place of religion both in the minds of individuals and in the institutional order, *This may or may not be associated with secularization, but is independent of it* (Berger, 2014: ix).

I agree that this is not, strictly speaking, an example of methodological sloppiness as such; it's not one of Berger's propositions, but the material I have italicized presents us with a statistical and practical impossibility. After all, it makes no sense to claim that we are uncertain as to whether two things are associated, but to simultaneously express certainty that they *are* independent.

Though he has long been a sociological tourist, Berger has been around sociologists of all methodological stripes long enough to know that two things can't be both associated and independent. This sort off statistical slip is easily forgiven in one who has taught us as much over the years as Berger has, all the while relying almost exclusively on the interpretative power of labor-intensive sociological tourism.

However, when Berger introduces the Kuznets curve, a simple economic model that purports to describe changing levels of income inequality in societies as they emerge from the most underdeveloped to the most modernized, substance rather than method becomes the problem. Referring back to Himmelfarb's *The Idea of Poverty* (Himmelfarb, 1984), Berger raises what amounts to a question of values: To which do we give priority, equality or productivity?

Simon Kuznets, who developed his famous and widely used curve, won a Nobel Prize in economics. While this does not constitute an endorsement of the excellence of everything he ever did, in a broad-brush fashion and over variable periods of time, the Kuznets curve, at first blush, seems a reasonable approximation of the changing character of income inequality in response to modernization as measured by increases in income per capita (Kuznets, 1968).

Berger refers to the Kuznets curve in making the case that economic development in capitalist societies typically starts off with very little income inequality and low per capita income, then rises rather sharply, meaning greater income inequality as mean per capita income increases, and then drops off to what appear to be really quite equitable levels of income inequality and shared substantial incomes. Berger's point is that in modernized societies capitalism is an economic system from which everyone benefits, without extremes of wealth and poverty (Berger, 1986).

However, while the Kuznets curve is disarmingly simple and has the look of plausibility, its use has been characterized by controversy, and it is not nearly as uncritically accepted as Berger claims (Fields, 2001). Berger's use of the Kuznets curve to measure equity as opposed to productivity is overdone and, in my view, has at best a dubious place in his discussion.

Berger as a Himmelfarbian

If there were such a creature as a Himmelfarbian, Berger would be one. He's committed to the notion that poverty is an inherently relative idea. I suppose this is what one would expect from someone who has never had an empty belly or slept in an unheated room in the middle of a cold winter, or sat for eight hours with an incarcerated inguinal hernia in a crowded hospital's emergency room. If Berger has experienced anything like this, he's forgotten about it. In any case, given his take on poverty and inequality more generally, why is he bothering with the Kuznets Curve?

In my view Berger makes entirely too much of the presumably intense differences between those who prefer equality and those who prefer higher levels of productivity (Berger, 1986). In my experience, the more important point has always been how low does the lowest stratum descend?

Rather than focus on the difference between the super-rich and those on the statistical floor of an income distribution, we would do well to acknowledge the material reality of poverty: inadequate nutrition that gives rise to disease, housing that does not support good health or no housing at all, lack of access to comfortable and protective clothing that facilitates playing whatever role one has been assigned, assuming one has a role and a social place, health care that does not meet one's needs and may gratuitously give way to death… (Iceland, 2003). We could continue, but the point is that there is nothing mysterious about poverty and what is needed to shed the miserable trappings of need.

In John Schwarz's exemplary but too frequently overlooked book, *Illusions of Opportunity* (Schwarz, 1998), he very effectively cites James Madison as claiming that one of the requirements of a durable society that invokes genuine allegiance is "economic competency." This simply means the existence of readily available opportunities to make a decent living for one's self and one's family.

Madison didn't concern himself with productivity versus equality at the national level, nor did he anguish over identifying and enumerating the constituents of a decent living. Some things are self-evident, and the difference between poverty and a decent living is one of them.

More of the pernicious influence of Himmelfarb

Rather than acknowledge this, Berger takes us on a Himmelfarbian tour of irrelevant arguments over the comparative desirability of equality and productivity. This may be pertinent to the work of professional economists, but in the context provided by *The Capitalist Revolution*, it is a specious diversion with the underhanded aim of directing attention away from real material and social deficiencies; a clumsy effort to explain away poverty so that we can simply ignore it (Abramsky, 2014).

If the capitalist revolution left many behind, so be it, or so Berger seems to be saying. Over the course of his career, he has indeed taken a hard right turn.

Yes, there are social classes, and their existence is durable and consequential (Wright, 2015). It seems that they will always be with us. But this fact does not provide an excuse for a mendacious argument to trivialize desperation. There's nothing relative about a shrunken stomach, abscessed teeth agonizingly removed with one's fingers, or living in a subway tunnel (Liebow, 1995).

In the process of finding modernizing virtue in East Asia, however, Berger quickly, almost surreptitiously, and using very few words dismisses less successful Latin American capitalist-engendered efforts at development with qualifying quotation marks ("capitalism"), sometimes, as in the case of Brazil, occurring under military dictatorships (Berger, 2011). Elsewhere, however, Berger takes the position that democracy is not a prerequisite for capitalist development, and that it is quite compatible with autocratic regimes and dictatorial regimes (Berger, 1986).

Capitalism, as Berger sees it, may be necessary for democracy, but democracy is not necessary for capitalism (Berger, 1986 & 2011). The dramatically rapid economic modernization of much of mainland China is offered as a case in point: nominally communist but in fact ferociously capitalist, and under the autocratic control of the Communist Party (Berger, 2011).

There is something entirely too glib and convenient about Berger's selective focus on East Asia and his rejection as not pertinent of cases inconsistent with his pro-capitalist thesis. Nevertheless, Berger argues that the perceived pain and loss of deeply internalized taken-for-granted meanings that accompany capitalist development are minimized when compared to any socialist or other alternative. Ever the Weberian, Berger takes the position that not all cultures are equally consistent with the requirements of development, and those that, in one way or another, emulate Weber's Protestant Ethic are in the best position to become prosperous through capitalist modernization (Berger, 1986, 2011 & 2014).

Functional equivalents of the Protestant Ethic

Though Berger, as we have noted, denies that he is a cultural determinist, he attributes East Asian developmental successes, in good part, to a shared post-Confucian ethos that he views as the functional equivalent of the Protestant Ethic. In contrast to classical Confucianism, governed by sacred ritual and the Mandarin hierarchy, in ways that Berger does not fully explain, post-Confucianism came to include an emphasis on the virtue of disciplined hard work as an end in itself (Berger, 2011).

Perhaps Berger is right. After all, he's not the first observer to make this kind of argument (see, for example, Kahn, 1979). It's hard to know, however, especially since the references Berger cites cover work done by himself or his collaborators. When Berger's methodological informality is extended to his review and acknowledgment of literature pertinent to his thesis, it ceases to be refreshing, and instead constitutes a self-serving obstacle to a persuasive presentation.

In addition, some scholars have judged Berger's post-Confucian thesis to be a gross oversimplification, one that applies Weberian interpretations to social systems that have so little in common with societies referred to by Weber that efforts to account for their functioning in Weberian terms make little or no sense (Tu, 1979/1996). In truth, Berger's later work is often inexplicably short on persuasiveness and scholarly discipline. One can only wonder if this is due to his self-proclaimed certainty as to the superiority of capitalism as an engine of economic development, or if it's a manifestation of generalized arrogance and the understandable certitude that comes with a lifetime of celebrated attainments.

Berger's book *A Far Glory* (Berger, 1992) contains some of the most beautifully written passages he has produced. However, it is also clear that much of the material in this fine book is based on informed speculation. Earlier we had noted that when Berger was reduced to speculation, he did not do very well, and our assessment of that set of circumstances has not changed. In the present instance, the material that has an ethnographic or historical basis is far and away the best written. Perhaps that is because Berger has been immersed in the sociology of religion for decades, thinking and writing about it in ways that enable him to render it masterfully. Perhaps his affinity for modernization and modernity is not as thorough and durable as his interest in religion. This could explain why the best prose in *The Homeless Mind* (Berger, Berger & Kellner, 1973) or *The Capitalist Revolution* (Berger, 1986) cannot match that in *A Far Glory* (Berger, 1992) or *In Praise of Doubt* (Berger & Zijderveld, 2009).

Nevertheless, it is difficult to take issue with Berger's judgment that there is more than one route to modernization and modernity, or his invocation of the notion that there are "multiple modernities" (Berger, 2011, Berger, 2014). Unfortunately, Berger leaves these observations, too, undeveloped and a good deal less informative than one would expect of the first-rate scholar that Berger long has been.

When Berger applies his Weberian thesis to countries as different as Japan, Korea, Indonesia, Thailand, China, most of the rest of East Asia, and later Jamaica, his inferences begin to seem suspect. As one among many cultural factors in each of these enormously diverse places, the very existence, much less the developmental efficacy, of something that can legitimately be compared to the Protestant Ethic requires some really torturously convoluted reasoning. I think that one could be excused for speculating that Berger is less interested in examining the continuing explanatory power of Max Weber's Protestant Ethic than in making a case for capitalism.

When Berger acknowledged that he is no longer capable of being even-handed when comparing capitalist developmental efforts with those geared to a socialist ethos, he willfully lost a great deal of credibility. He also overlooked the fact that capitalist and socialist need not be the only options, bringing to mind the modernizers, as well as differences within the capitalist and socialist categories.

Berger has observed more than once that "sociology [is] radical in its debunking perspective but conservative in its practical implications" (Berger 1963 & 2011). Fair enough. But is Berger the sociologist still capable, as needed, of debunking capitalist endeavors, or will he invariably give them a pass, somewhere finding a strand from the Protestant Ethic he can grab on to?

The Capitalist Revolution and *The Managerial Revolution*

Berger sometimes has trouble staying on topic, even when the topic is covered by the title of an entire book. In the present instance, as best I can determine, he strays because he wants to apprise us of a little-known, largely ill-considered threat to capitalism. I think that the characterization "strays" still applies because the threat is so implausible that acknowledging it inevitably takes us off topic.

Specifically, once, decades ago, there was a former Trotskyite turned right-wing public intellectual named James Burnham. He was a frequent contributor to William F. Buckley's conservative magazine *The National Review*, for which he wrote a regular column titled "World War III." All this would have nothing to do with Berger and his views on capitalism or anything else of interest to readers of *The Capitalist Revolution*, except that Berger has a long memory for obscure and pretty well discredited or simply ignored details that he inexplicably takes quite seriously.

In 1941 Burnham published a book titled *The Managerial Revolution* (Burnham, 1941/1972) in which he argued that capitalism was being subverted by a process that some clumsily referred to as managerialism (Enteman, 1993). As Burnham understood this process, it meant the separation of ownership from control in the familial and corporate worlds of great wealth and power (Burnham, 1941/1972).

Burnham assumed that managers, those who were gaining bureaucratic control, had interests that were often not identical with those of owners. As a result, a very crude but discernible approximation to socialized business practices and outcomes was becoming increasingly commonplace in the United States and elsewhere.

Berger characterizes Burnham's book as extremely influential, but I know of no one in the social sciences who takes the book seriously. Most, I would comfortably surmise, have never heard of it.

Even when it was first published, it was regarded as an oddity, written by a recently reformed Marxist with a theoretical axe to grind. It's similarity to a book titled *La Bureaucratisaiton du Monde* by an Italian theorist named Bruno Rizzi added probably unjustified cynicism to critics' and others' reception of Burnham's volume.

Burnham's position and that of his few remaining acolytes has always seemed to me to be a thoroughly implausible interpretation of developments in American capitalism in any of its periods, including the most modern. Ownership, after all, has prerogatives that can be transferred to managers and others only so long as the owner wants it that way. Owners' who are dissatisfied with managers' performance on *any* grounds can fire and replace them without giving the matter a second thought. Those who argue that owners lack the expertise of managers have only to look around to see an abundance of managerial experts ready to transfer their skills from one enterprise to another. Surely, owners can find managers who are tractable and who share the same objectives and outlook.

Besides, when managers are doing well, the owners on whose behalf they are working are doing well, too. This seems to be the vital connection between managers and owners that undercuts Burnham's argument. Recall that we live in a class-based society, and the feature of class that speaks loudest is money.

Perhaps because Weber had concluded that the class of career technical experts in managerial positions in government got their way more often than ministerial heads with greater authority, Berger may have seen substantial credibility in Burnham's thesis (Roth & Wittich, 1978). If so, ever the Weberian, Berger may have discounted all evidence to the contrary and subscribed to Burnham's managerialism.

The foregoing account includes a good deal of conjecture, but it's consistent with Berger's position that being a true-blue capitalist is not something for which a social scientist need apologize. After all, if Burnham is right, the concentration of ownership in the hands of a few is less perniciously anti-democratic and more consistent with the existence of a society where freedom prevails, independent of plutocratic influences. If managers are not pursuing precisely the same goals in precisely the same ways as would capitalist owners, the adverse influence of concentration of ownership is effectively diluted. Sounds like something we should celebrate, unless, of course, we subscribe to capitalism in a very specific form.

Berger, apparently one of the few who has taken Burnham's argument to heart, has speculated that it could very well yield a momentous modification in the American social stratification system, with those at the very top being managers, the ones in control, and those whose primary claim is ownership slipping beneath them. Nothing like this has happened, nor is there any reason to believe that this sort of shift is pending, even among organizations for which ownership is manifest in publically traded stocks (Domhoff, 2013).

Except for its vague and tenuous connection to Weber, Berger's acceptance of this preposterous characterization of the ostensibly changing nature of capitalism eludes understanding. With the controlling positions of entrepreneurs being confiscated by organization men, would capitalism still have the same institutional character? Would Berger still define capitalism as "production for a market by enterprising individuals or combines with the purpose of making a profit"? If something like Burnham's managerial revolution had, indeed, occurred or if it were in the works, Berger's simple definition would no longer fit quite so well, and perhaps that's what bothers him: he likes capitalism just as it is.

As with many others, Burnham emphasized the bureaucratization of modern society. We discussed this process at length with regard to processes of modernization, and what we said then still holds. Now, however, Berger argues that bureaucratization gives rise to a new type of person, someone less inclined to be favorably disposed to an intense entrepreneurial spirit, and with a different outlook and different set of skills. Participation in markets is still the essence of capitalism, but those imbued with the ethos of bureaucracy compete in different ways, an idea that, like so many others, Berger does not develop.

He does, however, claim the following:

> It can be plausibly argued that the decline of entrepreneurial 'individualism' and its widespread replacement by a managerial-bureaucratic ethos has created a threat to the survival of a capitalist economy and has made a socialist future far more probable (Berger, 1986: 108).

In my view, this is simply nonsense, that takes much too seriously the notions of enterprising individuals, as well as bureaucratic suffocation.

In truth, I think that Berger's use of the ideas of *The Managerial Revolution* in *The Capitalist Revolution* is a mistake of the sort that robs Berger's work of coherence and lends nothing of value to his presentation. Berger takes the perspective of the Burnham of 1942 as if it were gospel and fails to see recent developments. It's best, I think, that James Burnham remain long forgotten. See, for example, Berger's untestable hypotheses P6:

> P6: Certain components of Western bourgeois culture, notably those of active, rational, innovativeness, and self-discipline, are prerequisites of successful capitalist development anywhere (Berger, 1986: 109).

No doubt, but these same characteristics would serve one well in any bureaucratic organization where I've worked, and again see Blau on bureaucratic innovation (Blau, 1964). It's painful to see a first-rate scholar taking caricatures of innately dilatory bureaucracy as presented by tendentious others, and ignoring and failing to take concrete and dynamic adjustments seriously.

Nevertheless, others, notably Miriam Chandler (1977) have written effectively on what some call managerialism. In truth, the topic at hand is the organization of the modern corporation, an essential bureaucratic organizational form of big business, sometimes known as Big Capital. Approached specifically in that way, the literature is informative and timely. It does not, however, salvage Burnham's plodding misadventure, nor does it foretell the doom of capitalism. An apostle of Weber should immediately understand that bureaucratization and capitalism are not incompatible. Romanticized individualism does not undercut that claim.

Social mobility

The bureaucratization of modernized societies contributes abundantly to defining the changing processes of social mobility. In a world where self-employment and solo practice are increasingly difficult to attain, even members of traditional professions, medicine and the law, are often employed by bureaucratic organizations. For some aspirants, especially in the Third World, just gaining a white-collar position in a large bureaucracy may, in itself, be a significant step up.

Those of us who have spent a large part of our working lives as employees in bureaucratic organizations can attest to the fact that promotions may come very slowly, often depending on someone above us leaving without his position then being eliminated as a cost-saving measure. Moreover, on the odd occasion when promotions are dispensed, the large number who are passed over, already uncomfortable, are often left scratching their heads wondering how that person was chosen over so many who seem more worthy (Habermas, 1975).

Nevertheless, social mobility within a bureaucratic framework is a hard fact of modernized life. At the very least, it has the virtue of enabling one to follow, keep track of, and document upward and downward mobility, a real advantage should someone be doing research on the distribution of opportunities (Glassman, Swatos & Rosen, 1987).

Complementing our emphasis on social mobility as it occurs in bureaucratic settings where large numbers are employed Berger holds that "social mobility is basically driven by the industrialization process" (Berger, 1986: 59). There is no contradiction or inconsistency in juxtaposing claims of bureaucracy and industrialization in facilitating, or for that matter hindering social mobility. Instead, Berger is one of the few high-profile observers and theorists to recognize that social mobility driven by industrialization is, at least in part, a consequence of changes in the occupational distribution. Anyone who has examined the mobility tables that appear early in *The American Occupational Structure* (Blau & Duncan, 1967) will have seen this graphically portrayed using a large data set covering Whites in the United States at a time during the era of post-World War II prosperity.

Upward structural mobility

Blau & Duncan's (1967) analysis makes it abundantly clear that when the proportion of higher-level positions increases, personnel from lower levels will be pulled up in response to demand to fill newly created openings. This process is upward structural mobility.

By contrast, should the proportion of lower level positions increase, personnel from upper levels will be pulled downward. This process is downward structural mobility. Very generally, those who are pulled up are the better educated, and those pulled down are the less well educated.

You may recall, moreover, that in our discussion of human capital theory we objected to the implied but unmistakable claim that if prospective investors came across an area heavy-laden with human capital they would invest, expecting returns comparable to those gained from investing in any other form of capital. Investors thereby make an informed decision, fully expecting reasonable payoffs for investments in education (Becker, 1964/1993). The investments, ostensibly, create opportunities for upward mobility. Should the investments be withdrawn, the gains made by the well-educated will be lost to downward mobility.

However, it made, and still makes, a good deal more sense for us to reject this sort of supply-side model in favor of one that emphasizes demand (Robinson, 2013). In other words, social mobility and attendant remuneration for human capital are most promising when there are lucrative positions to be filled, and those who work for a living are drawn up to fill those positions. Berger neatly summarizes the joint operation of the demand-side and supply-side models in question form as follows:

> Are there jobs in the modernized sector of the economy? Are there opportunities for education and training making these jobs available to the individual? Does the individual have the ability and motivation to make use of the available opportunities (Berger, 1986: 63)?

The processes just described are driven largely by technological changes in the production process, placing a premium on innovation. This is the set of circumstances that prevailed in the post-World War II United States. Like all good things, it ceased to exist in this salubrious way when the number of available aspirants equaled and then exceeded the number of desirable positions (Newman, 1999; Thurow, 1980/2001; Uchitelle, 2007).

In consequence, Berger offers his next proposition, again ostensibly testable, regarding the nature of modernized capitalism:

> P7: Ongoing industrialization, regardless of its sociopolitical organization, is the basic determinant of social mobility (Berger, 1992: 60)

Once more, the claim that a proposition is formally testable is undercut by the absence of a specific basis for comparison: "ongoing industrialization" as opposed precisely to what? Nevertheless, it's easy to see what Berger is trying to explain, and with the words "regardless of its sociopolitical organization" he makes clear that he is not focusing exclusively on capitalism. Instead, he would expect industrialization to produce a similar result in a socialist society or in a hybrid cobbled together by modernizers.

With regard to the United States, this point of view, while persuasive, is becoming a bit threadbare. During the first decade and a half of the twenty-first century, finance capitalism has grown rapidly, transforming the once thoroughly industrial nature of the American economy (Hayward [Ed.], 2012). This is the sort of development that has prompted Krugman (2009) to warn that when finance capital gained ascendance in the past it has had extremely adverse consequences, including the Great Depression.

Berger does not discuss finance capitalism. It had not yet become nearly so important when *The Capitalist Revolution* was published (Berger, 1986). As a result, social mobility in circumstances where finance capital prevails is not addressed. However, to borrow an idea from Veblen (1904/2013), it's difficult to see how technological innovation in the production process can be accomplished by juggling pieces of paper, such as under-capitalized stocks, of dubious value. The role of finance capital in the industrialization process is indirect, and in itself it produces nothing.

Rates of mobility

At the time that Berger wrote *The Capitalist Revolution* (Berger, 1986), the following proposition made a lot of sense:

> P8: In all advanced industrial societies there have been moderate increases, but no dramatic changes, in the rates of upward mobility (Berger, 1986: 60).

However, social facts are no more immune to change than are natural phenomena that have an existence independent of social creation. As we have noted before, Berger is to be congratulated on not overplaying the rate of upward social mobility in the United States. As has always been the case, we are, in good part, creatures of the circumstances into which were born.

Furthermore, the nearly thirty years since publication of *The Capitalist Revolution* have seen damaging erosion in American's prospects for upward social mobility. Due largely to technological changes and the internationalization of capital intended for investment in job creation, prospects for upward social mobility have diminished, while downward mobility is an increasing common and demoralizing fact of economic and social life (Putnam, 2015).

One wonders at the long-term consequences of a debilitating change such as this. If poor prospects go unrelieved, a powerful source of legitimacy for

American society has been lost. In James Madison's words, "economic competency" for individuals and families is no longer available even for intelligent, diligent, and hardworking citizens (Schwarz, 1998).

Under such circumstances, Madison doubted that a society could long survive. An established social system may or may not have a moral obligation to assure livable economic opportunity for its citizens. But Madison held that as a practical matter the presence or absence of a moral obligation made no difference. In the absence of reasonably plentiful economic opportunity, social systems fail (Schwarz, 1998).

Economic competency, then, may be considered an important constituent of a capitalist society's economic culture, and the same would hold for prevailing payoffs and penalties for investment in education in the hope of using educational credentials as means to a better economic future. Each of these phenomena represents a crucial aspect of the circumstances in which capitalism is located. In light of our discussion of economic competency as an essentially important aspect of a society's economic culture, it is easy to see why Berger would be especially interested in upward social mobility.

Clearly, as we have repeatedly illustrated, the generation of testable hypotheses is not one of Berger's strengths. A little editorial assistance would have gone a long way toward substantial improvements in his presentation. Nevertheless, if we once again look past testability issues in an effort to make some substantive sense of P8, we see striking evidence that a comparatively brief period can make substantial differences in the way in which a modernized capitalist society functions, with corresponding effects on its citizens. Berger's tentative assertions or, if we prefer, propositions, including P8, in spite of their resistance to testability, still made a good deal of historical and social sense when *The Capitalist Revolution* was first published (Berger, 1986). That is no longer true.

In the United States, as we have already noted, continuing modest gains in rates of upward mobility are a thing of the past. Moreover, the once unthinkable may have happened: Downward mobility may have surpassed upward mobility. Recall that Berger rightly observed that patterns of mobility were due in substantial part to changes in the occupational distribution, meaning changes in demand (Berger, 1986).

At present, insofar as new jobs are being created, many more than in decades past are of the low-wage/no benefit service sector kind, while the sort of jobs that most aspirants would find attractive are more and more difficult to find because they have come to constitute a smaller proportion of the occupational distribution. This is overwhelmingly evident when one examines employment projections produced by the U.S. Bureau of Labor Statistics (Bureau of Labor Statistics, 2015).

Education and social mobility

This kind of information becomes even more troubling when an updated and properly interpreted response to Berger's next proposition is offered:

> P9: In all advanced industrial societies education has become the single most important vehicle of upward mobility.

The methodological difficulties with this ostensibly testable proposition are of the same kind that we found with its predecessors. Some of the terminology could be clarified or different words used, as with "vehicle," but I think the meaning is sufficiently clear that this hypothesis would be testable but, once again, for the crucial lack of something specific to which we could compare the efficacy of education. And testing an hypothesis that applies to "all advanced industrial societies" presents us with a truly monumental task. Nevertheless, we can see Berger's point easily enough.

However, Berger has already told us that ongoing industrialization is the fundamental source of social mobility generally, both upward and downward, and that one reason for this is because industrialization brings with it changes in the occupational distribution (Berger, 1986). If the changes yield an expansion of the upper levels of the distribution, participants will be pulled upward, and upward social mobility will predominate. So what is the role of education in this process? Why has Berger assigned it the status of the most powerful agency of upward social mobility? Perhaps the best thing we can say is that Berger's material is badly dated or, to be fair, "there are few who would not follow his example by assigning education the leading role" (cf. Beeghley, 2007).

Insofar as we have a problem, it inheres in the distinction between structural mobility and individual mobility (Blau & Duncan, 1967; Thurow, 1980/2001). If upward structural mobility is occurring it means that the number of higher level positions is increasing while the relative number of lower level positions is decreasing. As already explained, this will result in a comparatively large number of participants being pulled upward rather than downward. Fundamentally, this occurs without regard to the characteristics of participants: There are positions to be filled and someone must fill them.

Nevertheless, since not everyone, as a practical matter, will be pulled upward, those who are will usually be the best educated. The less well educated are more likely to be left behind. However, the role of differences among individuals with regard to education or anything else is secondary in the process of upward structural mobility. The primary factor is the changing shape of the occupational distribution (Berger, 1986).

Notice, moreover, that the kind of rationality that Berger invokes in his propositions is Zweckrationality, based on means-ends relationships. Wertrationality, or use value, is not in evidence.

Individual mobility: a zero-sum game

Imagine that the shape of the occupational distribution is not changing. The relative number of lower, higher, and middling positions is constant. This does not mean that there is no upward or downward mobility, but it does require that mobility have the character of a zero-sum game involving individual participants (Thurow, 1980/2001). Whenever someone moves up, someone else must move down. Moreover, the comparatively well-educated are more likely to move up while the others are more likely to move down. This kind of mobility is referred to as individual mobility in contrast to structural mobility.

In the real world that we share, both kinds, structural mobility and individual mobility, are going on at the same time. With respect to each, the better educated are at an advantage. However, we can see that payoffs for investments in education occur more frequently when upward structural mobility is occurring, with changes in the occupational distribution and investments in education working together to generate upward mobility (McShane, 2016).

The United States at the present time is not a place where upward mobility of either kind is predominant over downward mobility. Sadly, for the long term, relatively few new jobs are being created, and most of those created are in the lower levels of the distribution. At the same time, the labor market, where human capital is bought and sold, has been glutted with relatively well-educated people. Under these circumstances, with glutted labor markets and minimal job creation, upward mobility will be painfully infrequent while downward mobility will be much more common (Bowles, Edwards, & Roosevelt, 2005).

This represents a sharp departure from the post-World War II period of prosperity when upward mobility predominated and almost anyone could get a job that would enable them to earn a living. It's easy to see why this unfortunate development has occurred, but how do we respond? The tried and once-true method of investing in education, tacitly endorsed by Berger in *The Capitalist Revolution* (Berger, 1986), will more thoroughly glut the market for human capital with an even larger number of comparatively well-educated people. A difficult set of circumstances, indeed.

14 What has been done?

Fittingly, a compelling and informative point of this brief chapter inheres in the unmistakable importance of material prosperity as a consequence of modernization. Yes, modernization brings other sorts of outcomes as well, but Berger is first and foremost interested in material prosperity. I have no argument with this emphasis, but it tacitly raises once again the comparative value of materially well-endowed and spiritually rich sets of circumstances.

We are left, thus, with a question that has long defied a persuasive and honest answer: Which is more important the material or the spiritual? Berger gives no reasons for his choice of material prosperity, though in a modernizing world we can certainly think of some, and perhaps this is what Berger expects us to do.

It may be, however, that Berger's point is one that Marx would readily endorse: Nothing good comes of hunger, ignorance, and disease. Less obviously, the same applies to class anxiety at the prospect of downward mobility. Little wonder that the old middle class and the new middle class exist in a conflictual relationship.

This illustration contributes to establishing the centrality of conflict and the undeniable importance of material prosperity in understanding modernization and modernity. In spite of their importance they are easy to overlook, though *The Capitalist Revolution* brought them clearly to the fore. Nevertheless, in a discussion of modernization and modernity, they deserve the emphasis provided by one brief stand-alone chapter.

15 Styles of consciousness and pluralism

Berger wants to emphasize that a life plan is not only an essential part of the process whereby one's future is tentatively laid out and one's identity created, but that it is typically multi-relational. Life plans, in the modernized world, commonly include all the members of a nuclear family. As a result, multiple futures are being planned, some perhaps more advantageously than others. And multiple identities are being created, and again not all may be equally well served.

Identity formation

Identity formation will be a more complex process, one fraught with conflict and social complexity, in a highly pluralized society. There are simply too many components that could be organized in too many different ways for creation of an identity devoid of cracks, fissures, and schisms, terms that Poulantzas (1978/2015) used in capturing the difficulties involved in holding together the state in late capitalist society. For some reason, the analogy appeals to me, perhaps because in each instance the emphasis is on fragility.

Identify formation in a pluralized society is much less stable than the same process in a simpler social system. The more pluralized society is shot through with countless potentially conflicting loyalties, prospects, and ways of succeeding and failing. This is yet another reason for maintaining the family intact while constructing a life plan. In a modernized world the family is of paramount importance in maintaining understanding of one's self and one's place (Berger & Luckmann, 1966).

Identity formation in a highly pluralized society never ceases to be buffeted, beaten, and open to pernicious seduction by an ever-changing array of social forces, so it seems reasonable to take seriously the claim that it is never complete. Ironically, however, as Berger insightfully explains, a stable identity may be of even greater importance in a modernized society than a traditional one, because so many things in a modernized society are so often changing, presenting participants with a world that is heavy-laden with new and still-transitory meanings (Berger, Berger & Kellner, 1973).

As a result, if one's identify is not stable, then it seems that nothing is, and our lives are immersed in fundamentally meaningless chaos. The fact that each one is responsible for the identities of others, namely members of one's own family, makes this anomic burden even more difficult. Clearly, the existence of a multiplicity of overlapping relationships in a modernized society is, at best, a mixed blessing.

Even given the persuasiveness of this view, I still think that my objection to the clumsy ostentation of the neologism "multi-relational synchronization" is warranted (Berger, Berger & Kellner, 1973). It serves scientistic mystification better than explanation. Nevertheless, the fact that the modern family assures that any decision one makes for one's self may have a profound influence for others is undeniably true, and is cause for serious concern. How one should handle this is rarely unambiguously clear.

Religious faith

An important topic we have not discussed and that belongs in any account of social cohesion and stability in the modernization process is the institutional status and efficacy of religion. The role of religion as a cohesive force in traditional societies is well known (Durkheim, 1912/1995; McGuire, 2008). However, when pluralism brings us an abundance of religious choices, religion's cohesive force may be badly weakened. In good part, this may be due to the fact that any one set of religious definitions of reality is now subject to challenge by one or more others, and the differences are known to the public at large (Berger, 2011 & 2014).

Recall, moreover, that as a society becomes more rationalized, its institutional order is subject to ever closer scrutiny. Rationality and religious faith are often difficult to reconcile. This does not mean that the faith prescribed by any religion is false, just that when evaluated in terms of conventional functional rationality, it does not fare well.

It is important to remember, however, that functional rationality is not the only basis for evaluating religious faith (Bellah, Madsen, Sullivan & Swidler, 2007). Thus, we may ask a question similar to the one asked about morality: Does it make sense to ask about the veracity of religious truth once we have recognized that religious systems, too, are social constructions (Berger & Kellner, 1981: 84)? The best answer I can come up with is "Of course!" I'm sure that Berger would agree. Yes, we live in a modernized society, and rationality pervades all institutions and contributes a factor of overriding importance to prevailing styles of consciousness. In itself, however, rationality is without substance, and its ascendance leaves us, eventually, with a cognitive void.

This is a point made very effectively by Horkheimer & Adorno in *Dialectic of Enlightenment* (1947/2007). The mind has, indeed, become homeless and empty of durable substantive content, a painful set of circumstances rendering us vulnerable to sometimes misguided and self-defeating efforts to fill the void and find a home and a purpose. When Horkheimer and Adorno wrote, the concrete threat was fascism, but the sources of vulnerability were the same as those posed by modernization.

To again borrow concepts from Durkheim (1893/2014), we become open to egoism, or lack of membership, and anomie, or cultural deregulation. In circumstances such as this, the appeals of the most rigid fundamentalists and fanatics may become much more attractive.

In recognition that technology-intensive modernization is not without substantial and pervasively painful consequences, Berger frequently invokes Durkheim's well-known concept, anomie. When he does so, however, I think he uses it to cover both the cultural deregulation and attendant meaningless with respect to which Durkheim used anomie, as well as the closely related Durkheimian notion of egoism or social isolation devoid of any sense of membership.

Berger, no doubt, has his own reasons for this conceptual truncation, but its continued use is worth pointing out. The distinction between the two concepts was important to Durkheim (1893/2014), and it should be at least as important to students of modernization as a process that has both favorable and unfavorable outcomes, an issue to which Berger is keenly attuned. Playing fast and loose with established concepts serves no one's interest, least of all the person doing the playing.

I've decided to let the foregoing material on religious faith stand even though I'm having second thoughts about rationality and religious faith. Having re-read sections of *Sociology Reinterpreted* (Berger & Kellner, 1981), I now realize that if I were sufficiently knowledgeable it would be entirely possible to adopt an historically informed relevance structure that could make Catholicism and perhaps any other religious faith interpretably rational.

Perhaps the qualifier "interpretably" renders this addendum less compelling, but the fact remains that if I knew enough about the history of Catholicism I'm sure I could creatively but honestly interpret the information to provide a substantively persuasive foundation for Catholicism. Tendentious or not, at face value it would stand up to the test of rationality. Looking deeper it wouldn't do so well, but who knows? It still might be true.

True or not, the persuasiveness of any one religious commitment in most societies has long been much easier to maintain than today because most societies had only one (Durkheim, 1912/1995). Pluralism, thus, was not a problem for the faithful or their leaders. The truth of this statement is brilliantly captured by Berger in the following quotation:

> Every human society has institutions and functionaries whose task it is to represent this [putative] religious truth, to transmit it to each new generation, to engage in rituals that reaffirm it, and sometimes to deal … with those who are benighted or wicked enough to deny it. In most societies in history this has been a relatively easy matter because there was only one set of beliefs and values, a unified and coherent worldview that everybody knew and almost everybody took for granted (Berger, 1992: 11).

When there is more than one faith, each with a plausibility structure of righteous adherents, concepts such as fundamentalism and relativity, fanaticism and tolerance are invoked, perhaps for the first time, to explain the prevailing set of circumstances. Much the same is true for Berger's graphically descriptive expression "reactive counter-formation." Berger uses this concept to provide a concise and carefully targeted explanation of what the religious movements represented by contemporary Islam and Evangelical Protestantism have in common (Berger, 1992: 33). Most instructively, they are reactions to the discomfort and dislocation caused by modernity. For those who have read *The Homeless Mind* (Berger, Berger & Kellner, 1973) and understand the meaning and content of the symbolic universe, this will be easy to understand.

Both movements, moreover, run counter to secularization theory. Given the breadth of the two revitalizing religions, we can imagine that they force secularization still more tightly into Western Europe, with perhaps a more modest expansion of secular forces into Southern and Eastern Europe. However, while Islam is socially reactionary, re-shaping the context in which it occurs into a recrudescence of the distant past, Berger finds Evangelical Pentecostalism replete with modernizing influences. As is often the case with Berger, he finds in the abstemious, hard-working, future-oriented character of the Pentecostals much of the virtue he tirelessly attributes to the Protestant Ethic or something very much like it (Berger, 1967).

Certainly Western Europe is a thoroughly modern region. If one were to rely on Western European countries for evidence that secularization theory has merit, he might very well conclude that it does. However, the United States is also thoroughly modern, but its level of religiosity and other forms of diversity have been comparatively high and perhaps increasing, a development that is sharply at odds with secularization theory.

Berger takes the contrast between the two areas to mean that, on balance, modernity is not a necessary and sufficient condition for secularization. Though he never puts it in just these words, I think Berger's position is best summarized by judging that a combination of pluralism and modernity provides the best foundation for the ongoing development of secularization (Berger, 1992: 37). The rate of secularization in the United States, moreover, is further diminished among churches that are relatively conservative, adhering to time-honored beliefs and practices.

If the United States had proportionately fewer well-educated, affluent, upper middle class citizens than Western Europe, we would have a plausible though still very tentative explanation for its lower level of secularization. However, in terms of the factors just mentioned, Western Europe and the United States long have been roughly the same (Berger, 1992).

16 Reservations about pluralism

Furthermore, Berger seems to have concluded that the sources of pluralism are introduced from the outside, as a modernized civilization makes incursions into a primitive society, There is no doubt a great deal of merit to that claim. However, I see no reason to believe that a society, however primitive, can be totally devoid of eccentricities, at odds with the norm. The further development of these would surely give rise to pluralism. Much the same may be true of internal variability in the strength and weakness of established institutions, giving rise to natural developments that create things and ideas unimagined before. I am not rejecting pluralism as Berger understands it, simply saying that its existence has long been recognized by some of his most brilliant predecessors and saying that its sources are sometimes – perhaps – internal.

At this point I will make an admission that will, for better or worse, qualifies much that I have written or will write in this manuscript. I have to admit that the more I refer to pluralism and its secularizing consequences, the more I come to suspect that Durkheim (1897/1997 & 1912/1995) and others had captured pluralism in their own accounts of secularism, they just didn't explicitly use equivalent terms for "pluralism" and "secularism" in whatever language they were writing.

Diversity, after all, however it came to be, betokens a proliferation of choices, whether consciously made or not. Choices, moreover, are by their very nature made among a plurality of entities or phenomena, whatever their source. I think that Berger's discovery of pluralism does give a different slant to more customary ways of explaining secularism, but as a theoretical innovation its contribution is more linguistic than substantive.

Berger seems to have oversold himself on the explanatory power of pluralism. I think it's understandable that when a sociologist or other scholar comes up with what seems to be an unexpectedly productive insight, he or she will embrace it to the exclusion of alternatives, and that may be what Berger has done by substituting pluralism for secularism, or at least treating secularism as a practically inevitable consequence of pluralism (Berger, 2014). This, however, is just more or less plausible speculation on my part.

It becomes quite clear, however, that Berger is not just playing with ideas and idly offering alternatives to prevailing points of view when he makes the unqualified assertion that "Modernity does not necessarily secularize; it necessarily pluralizes" (Berger, 2011). Thus a point of view that was entirely foreign to me, and one that I find myself resisting the efficacy that Berger assigns it, is made intrinsic to his influential work on modernity and modernization.

It is important to acknowledge that Berger's seemingly intransigent commitment to pluralism and its secularizing power is not so stubbornly manifest that it denies all evidence to the contrary. This is especially evident in *A Far Glory* (1992), in which Berger unambiguously holds that pluralism and modernity both play a substantial role in secularizing large parts of our world. The acknowledgment is welcome, and it diminishes the force of my objection to what I had taken to be an overemphasis on pluralism. This more temperate view of pluralism is manifest in the following definition (Berger, 1992: 86):

> Pluralism denotes a state of affairs on which no single group constitutes the society as a whole and, in consequence, no group can serve as an all-embracing community for its members.

While this definition is a bit oblique, not really directly addressing the "plural" in "pluralism" and not invoking the concept secularism, it has the virtue of avoiding exaggerated claims as to the effects of pluralism. As such, it gives us a way to think about pluralism that is both substantively useful and suitably understated

Modernization and the once taken-for-granted

Going still further in his insertion of pluralism in the place of secularism, Berger's position is that modernization diminishes the cultural force of what once were taken-for-granted preconceptions and beliefs, but not because modernization makes the world more secular. Instead, the diminution and demise of the taken-for-granted that once characterized a specific, unitary world view occurs because of pluralism. Beliefs that had been regarded as undeniably sacred, embodying the only possible way to believe and worship, are now beset by alternatives. People, for the first time, are given choices with regard to what they believe and the world view they entertain. The choices, moreover, are rendered inescapable by mass media and the frenetic complexity of an information-laden urban environment.

At first glance, this seems entirely too glib and convenient. Strongly held beliefs and perspectives that have been with one for a lifetime seem very unlikely to be traded in just because alternatives become available. However, Berger makes his positon a good deal more solid when he adds that modernization doesn't merely introduce alternatives, it also diminishes the integrity and grasp of the once tightly-knit communities that provided the social foundation that supported and maintained the sanctity of traditional beliefs. As a result, alternative judgments as to what is sacred and, more generally, as to what is real are no longer unthinkable (Berger, 2002).

In *The Sacred Canopy* (Berger, 1967) Berger refers to these reality-maintaining groups and communities using the by now familiar but still ingenious term "plausibility structures," a neologism that in suitable context has a self-evident meaning. Dissolution of the structured collective renders once unquestioned beliefs for the first time suspect or a good deal less plausible. As is typically the case, activity, in this instance taking the form of group maintenance or lack of same, exists in a dialectical relationship with ideas, especially religious activities and ideas that are brought to the fore (Berger 1967). The importance of plausibility structures in the interplay among the sacred, the secular, and the plural is highlighted by their place and performance.

Internal eccentricities and vulnerability

It's certainly worth noticing, however, that the demise of cohesive community support leaves one not merely open to pluralism, but renders people diffusely vulnerable. This vulnerability is the sort of condition that may lead one to embrace eccentricities of uncertain origin that are intrinsic to the once cohesive community. Feeding on itself and its own uncertain intrinsic influences may give rise to previously unknown religious feelings and a rudimentary pattern of beliefs. There need be nothing pluralizing coming from the outside. It's enough that the community loses its cohesiveness – perhaps someone has discovered gold – and that the new-found absence of near-perfect internal homogeneity leaves the people open to miscellaneous influences (Tönnies, 1887/2001).

Berger produces a bit of confusion in *Between Relativism and Fundamentalism* (Berger [Ed.], 2009) when, to paraphrase, he raises the possibility of substantially increased secularism and pluralism due to the influence of a home-grown, world-class genius on the social system to which he is born, bred, and lives out his life. I find this example especially interesting because it suggests strong secularizing and pluralizing consequences due to internal developments, without external influences. The crucial question is whether this is in fact a pluralizing instance that would not have been understood as such by Durkheim (1883/1997), Tönnies (1887/2001), and other classical theorists.

In my view, this would be an example of nascent secularization without externally promulgated pluralism. If we introduce the external influences of modernization, pluralism may be manifest, but it need not have the social power of internal developments that may heavily overshadow the influence of external factors.

In any case, the reservations I expressed about the place of pluralism in the process of technology-intensive modernization still stand. I think that this is an idea that was productively implicit in the work of others, Durkheim (1883/1997) in particular, concerned with the transformation of the sacred into the secular. It is certainly true, moreover, that the two phenomena, pluralism and secularism, occur in the same set of social circumstances. This is something that Berger makes clear when he asserts that a better understanding of pluralism would enable us to understand why Europe, according to his reckoning, is more secularized than America (Berger, 2011). In fact, Berger holds that Western Europe is perhaps the only large area where secularization theory unequivocally holds sway (Berger, 1992).

Even then, however, Berger gives priority to pluralism, presenting it as a causal precursor of secularism. At the risk of redundancy, I think a more tenable positon as to the advent and eventual pervasiveness of secularism is something that acknowledges that diversity, however it came to be, betokens a proliferation of choices, whether consciously made or not. Choices, moreover, are made among a plurality of entities or phenomena, whatever their source, internal or external. After all, social entities of all kinds are subject to internal development. Berger's discovery of the idea of pluralism may give a different slant to more customary ways of explaining secularism, but as a theoretical innovation its contribution and magnitude are, I think, of uncertain value.

For better or worse, I have based the distinction between pluralism and some instances of secularism on external sources as opposed to internally prompted developments. My reading of Berger has convinced me of the decisive importance of external factors for pluralism in his perspective. The place of internal developments without prompting from the outside is a presumption of my own informed by Durkheim (1883/1997). Perhaps I have misconstrued Berger's use of pluralism as necessarily due to external forces when, in fact, it is not.

In any case, in spite of my reservations about pluralism expressed throughout this manuscript, I've found it easy to use the term pluralism in the manner Berger prescribed.

17 The complexity of modern consciousness

Marx's vision, in which abysmal poverty takes hold of the vast majority of the long-term capitalist populations, has not happened, unless we view things on a world scale, but in any one modernized society it is virtually unknown. Perhaps that will always be the case. Nevertheless, examine the distribution of wealth and income in the United States. The capitalist class does, indeed, enjoy a concentration of material resources, while many may suffer economic stagnation or downward economic mobility. Marx I think, understood the logic of capitalism far better than Berger dare acknowledge, and his expectation of eventual immiseraton is currently being roughly approximated.

In discussing life and conflict under conditions such as these, Marxists and those heavily influenced by Marx, often use the term "material practice." This is not part of Berger's active vocabulary, but he certainly understands and uses the term's conventional synonym, socialization, to refer to the creation of particular kinds of human beings whose characteristics are determined by concrete experience in the social world. That this process will yield different outcomes in response to different circumstances seems obvious, and Berger's discussion of the creation of distinctive selves in the modernized and modernizing worlds is illuminating.

In developing his understanding of modern consciousness and its institutional context Berger seems to struggle again. In *The Homeless Mind* (Berger, Berger & Kellner, 1973) he used borrowed concepts such as package and carrier to manage ideas that would have been more conveniently served by conventional usage (Berger, Berger & Kellner, 1973). Nevertheless, Berger apparently liked the new and different names for simple, old ideas, and he uses them again in his introduction to *Many Globalizations* (Berger & Huntington, 2002). As I understand him, Berger wants to clarify the relationships between styles of consciousness and the nature of institutions that are inevitably found in a modernized society, taking into consideration both their intrinsic (essential) and extrinsic (non-essential) properties.

In my view, the first and by far most important point is that our cognitive content and style of consciousness are products of what we do and the nature of the institutional settings where we do it. The way we think, in this view, is determined by our activities, and our activities are shaped by the institutionalized avenues of activity that are open to us (Berger & Huntington, 2002).

As we have seen, it is from Marx that the sociology of knowledge developed its fundamental proposition that man's consciousness is determined by his social being, something that Berger unambiguously acknowledged (Berger & Luckmann, 1966; also see Fromm, 1981). From Marx, moreover, as Berger makes emphatically clear, the sociology of knowledge inherited two of its most important concepts: ideology, or ideas fashioned to serve particular social interests, and false consciousness, or "thought as alienated from the real social being of the thinker" (Berger & Luckmann, 1966: 6). Berger's simple, brief, and astute explanations of ideology and false consciousness are examples of the author at his best.

Since Berger's work brings polemical praise to capitalism, and is in substantial part an antagonistic account of Marxism, it is useful to recall that Berger and Marx have a good deal in common with regard to how the content and flow of our thought processes are determined (Berger, 1986). That said, it merits noting that Berger's judgment that Marx's predictions of the immiseration of the working class and the sharp exaggeration of class differences were patently erroneous have, in my view, become problematic.

Immiseration revisited

In the United States especially, but in other Western industrialized democracies, as well, the distance in wealth and income between members of the capitalist class and just about all others has grown enormously, and continues to accelerate (Draut, 2007). Moreover, the last three decades have seen the totally diminished credibility of the too-convenient canard that a rising tide lifts all ships. Instead, while the incomes of the richest people and vastly wealthy corporations have been dramatically elevated, the rest of us have seen our economic wherewithal stagnate or decline (Pikitty, 2014).

Whether we attribute this to technological development, globalization, or something conceptually less sanitary, class differences have widened enormously and something much like what Marx meant by immiseration seems clearly to be taking place. In the 1960s and early 1970s we would have had to acknowledge that Berger's critique of Marx on this issue had substantial merit. As Berger has emphasized, however, capitalism moves rapidly, developing in ways that even the best and the brightest will not foresee.

Furthermore, the commonplace and well-informed characterization of the United States as an oligarchy may be on the way to attesting to Marx's astuteness when he held that we would eventually find "dictatorial regimes in the heartlands of capitalism" (Berger, 1986: 5). Political choice – democracy – in the United States has been subordinated to money. And large concentrations of money have come to narrowly limit not only those who run for office and are elected, but the increasingly unfavorable distribution of occupational opportunities and the cost of sometimes worthless credentials. Call this set of circumstances what you want, but if it's not yet dictatorial, it's certainly not democratic.

Material practice

In my view, as noted above, the first and by far most important point for the sociology of knowledge is that our cognitive content and style of consciousness are products of what Marxists sometimes refer to as material practice. The way we think, thus, is determined by what we do and what we do is shaped by the institutionalized avenues of activity that are open to us. In other words, what and how we think is determined by the character of our institutionalized social existence (Marx, 1867/2011; Bottomore, 1956). The frequent references to Bottomore's *Selected Writings in Sociology and Social Philosophy* (1956) throughout this manuscript reflect its sustained and remarkable value as a source of information on Marx.

The ideas subsumed by material practice are well known, and I find them convincing, again bringing to mind the notion that Berger is a philosophical materialist. Berger seems to accept this way of explaining socialization, but he is also convinced that the determination and nature of consciousness in the modernized world is more complex than a straightforward and simple materialist formulation prescribes.

Take, for example, the idea of componentiality, meaning, as we have seen, that anything produced can be assembled and disassembled into discrete parts, a property that Berger views as intrinsic to technology-intensive production, and something that he presents as inevitably manifest in modern consciousness. People themselves, moreover, are essentially components in the process of production, something not difficult for them to see and internalize.

This holds by virtue of the fact that technology-intensive organizations of production are ubiquitous in the modern world. We cannot avoid directly or indirectly experiencing them and their mode of organization and operation, nor can we escape their effects on our style of consciousness.

However, Berger wants to make clear that componentiality, at least in the United States, is not experienced and internalized purely and without extrinsic adornment. Instead, it is accompanied by properties "derived from the economic ethos of capitalism" – "for example the motives of individual competition and profit-making geared to the accumulation of private property" (Berger, Berger & Kellner, 1973: 122). This seems obvious enough, but Berger insists on making it explicit.

It is worth noting that a Marxist would give priority to capitalism over technology-intensive organizations of production (Marx & Engels, 1846/1998). However, at this juncture it matters little which perspective, Berger's or the Marxists', is correct. The more important issue is that institutional analysis is not a nakedly empirical activity, but presupposes an expressed or implied theoretical orientation. Berger has a point.

The complication introduced by the need for theoretical choices makes an additional contribution to affirming Berger's claim that institutional analysis, with a view to determining the role of any or all institutions in defining a style of consciousness would, indeed, be an extraordinarily complicated and difficult process. To his credit, moreover, Berger acknowledges the difference between institutional analysis done by a Marxist and that done by one who shares Berger's theoretical perspective.

As a proponent of classical social theory as rendered by Max Weber, Berger wants to show his readers that there is a reciprocal relationship between institutional structures and consciousness. Institutions, as we know are socially constructed by the cognitive content and style of consciousness of those who actively and consciously participate in them.

With regard to determining the nature of a modern style of consciousness, Berger, as one might anticipate, gives priority to technology-intensive production and the organizations that we have referred to as instances of political bureaucracy. For good or ill, Berger refers to these as primary carriers of the content and style of consciousness (Berger, Berger & Kellner, 1973).

As we have emphasized, however, the organizational forms that have given us modernization and modernity are also often sources of alienation, anomie, and egoism. They foster creation of a society where individuation is liberating, but at the same time promotes the sense of homelessness that is a recurring theme throughout processes of modernization and the modernized world itself.

Modernity, once that conceptually elusive and uncertainly defined state has been achieved, brings material abundance for many, but it is often accompanied by a profound sense of isolation and meaninglessness. No wonder that processes of counter-modernization and demodernization, of greater and lesser sophistication and efficacy, form to address the discontents of modernity.

18 The continuing influence of Weber

Nevertheless, the millions of readers of Berger's first book, *Invitation to Sociology*, may find in its first five chapters some of the most compelling accounts of social determination they will ever read. The chapters are truly brilliant expositions of the sociological perspective in it most persuasively deterministic form. Berger backs off as he makes a really awkward transition to an unbridled free will in the last three chapters, and these are probably best left unread.

Determinism is manifest in Berger's discussion of social roles and role relationships, concepts he discusses with impressive sure-handedness. In *Invitation to Sociology* and elsewhere, Berger makes as strong and temperate a case for social determinism that one is likely to find anywhere (Berger, 1963). To his credit Berger also clarifies the overwhelming difficulties involved in efforts to study free will itself, an explanatory endeavor that I have never seen before.

What we have presented so far gives us a clear view of how Berger understands institutional complexes, how they affect styles of consciousness, and how consciousness reflects back on institutional structures in a dialectical pattern of reciprocal determination. Ever the Weberian, this is the way Berger sees the social world working, and he succeeds in making the case that modernized consciousness is necessarily quite complex and may be fraught with discomfort.

Modernization has succeeded in what it does best: rationally employing technology-intensive production and bureaucratic rationality to foster material improvements and prosperity. In that limited sense, it has dramatically increased the number and quality of options open to those who live in a modernized world (Berger & Kellner, 1981). This might be interpreted to mean that modernization has increased human freedom, but I think that is vastly overstating the case, especially if we take increased human freedom to mean qualitative improvement in the way people live. Modernization has failed, after all, in providing a style of consciousness that is imbued with purpose, meaning, a sense of belonging, and the strong feeling that one has a home (Berger, Berger & Kellner, 1973).

Free will

Consistent with his favorite classical theoretical perspective, Berger rejects the notion that free will is chimerical, and that all behavior is socially determined. He acknowledges, however, that there are more than a few sociologists who take the fully deterministic positon quite seriously. He may have in mind sociologists influenced by structural Marxists, for whom human beings are props, more or less mindlessly supporting established structures of domination and exploitation (Althusser & Balibar, 1970). For Berger, however, it is essential that we acknowledge that socialization is never perfect or complete.

This is the same Berger, however, who reminds us that with regard to just about any situation we encounter, the game has been "fixed" long before we arrive on the scene. All that is left for us to do is to play it with more or less enthusiasm and skill (Berger, 1963: 86).

In any event, it is clear that determinism has its limits, though Berger may in some crucial respects exaggerate those constraints. Nevertheless, experience in the social world cannot, for example, socialize George Herbert Mead's spontaneous "I," which exists in dynamic tension with the thoroughly socialized "me" (Mead, 1934/1967). As Berger, Mead and many others see it, in the absence of free will there would be no way to account for originality or innovation of any degree or kind.

No doubt Berger has a point, but in making it in various venues he seems to be trying a bit too hard (cf. Berger & Kellner, 1981). In *Invitation to Sociology* (Berger, 1963), the first five chapters, comprising the bulk of the book, are quite consistent with a thoroughgoing social determinism. These are also the most compelling and well-written chapters in this truly fine sociological classic.

In the three subsequent chapters, however, Berger steps away from social determinism, and seemingly out of a sense of professional obligation, tries to open avenues for behavior that is not socially determined. This material is not nearly as persuasive and coherent as that which went before, and it sometimes has a histrionic feel. It is by no means the work of a sociological amateur, but it is a good deal less persuasive than most of Berger's work.

Some readers might understandably take the last chapters of *Invitation to Sociology* (1963) to be a prescription for ways to use sociological concepts as excuses to engage in exercises in bad faith. However, this transforms sociology into a much less deterministic discipline than it has ever seemed to me to be, one in which we can step in and out of all thoroughly internalized social roles as if they completely lacked holding power. Simultaneously, we ignore established institutional structures as we find it convenient, and as if, contrary to fact, they were devoid of objective facticity.

Social roles

We should bear in mind, moreover, that individuals experience the objective facticity of institutions by playing social roles. Roles are primary institutional constituents, and they are created through the same processes of externalization, objectification, and internalization as the institutions they constitute (Berger, 1967). We play roles competently only after we have been thoroughly socialized into their requirements in a well-informed way.

The requisite information is part of our common stock of knowledge that contains the reciprocal typifications constitutive of the nature of a role. Construing a role in this way makes clear that a role is a social phenomenon, and while individuals are socialized into roles and play them as such performances are called for, a role is best understood as a characteristic of a social situation, much as the roles of pitcher, catcher, batter, and first baseman constituted a social situation in our earlier baseball game example.

Institutions, in general, consist of collections of interrelated roles. Berger, however, may unintentionally mislead us when he asserts simply that "all institutionalized conduct involves roles" (Berger & Luckmann, 1966: 74). Berger rightly holds that in any institution some roles are more important than others, but beyond the distinction between primary and secondary socialization, he fails to adequately acknowledge that some roles entail a much more thorough and binding hold on their occupants than others.

As a result, at various places throughout this manuscript I have faulted Berger for taking the implied position that people can step in and out of roles as it suits them. My reason for doing so was that even secondary socialization into the requirements of a role entails obligations that may or may not be easily cast off. If secondary socialization has been successful, the moral grip that the role exerts on its occupant will be too strong to permit a role to be relinquished without a good deal of effort.

However, just as there are institutions that direct human social behavior with greater and lesser degrees of force, much the same applies to the roles that constitute them. It seems reasonable to argue that, in general, it would be much more difficult for a mother to abandon her infant children than it would be for a part-time, minimum-wage grocery store clerk to abandon his co-workers. The likely consequences are very different, the sense of loss may be intolerably different, and the processes of socialization, whatever form they may have taken, are much more thorough and binding for mothers than for grocery store clerks.

In short, it really is possible to easily step in and out of some social roles and to do so painlessly and with little effort. With regard to other roles, however, once adopted they may be nearly impossible to relinquish. Perhaps the social efficacy of an institution can be roughly gauged by the indifference or concern with which an entry-to-exit path can be traversed. But that's just speculation.

Sociology: seeing through or seeing?

In a roughly comparable way, I am uncomfortable with Berger's judgment that sociology is a way of seeing through prevailing social structures (Berger,

1963). There may be merit to this assertion, but I think it makes a lot more sense to realize that sociology enables us to see and give meaning to those structures in the first place. Insofar as sociology has the potential for liberation, it is by enabling us to at least discern the structures that determine our behavior (Berger & Kellner, 1981).

Perhaps these are just two sides of the same coin, but a taken-for-granted, commonsense understanding of phenomena such as social class is very different, and of much less analytical value than the same ideas as rendered by, for example, Marx (1844/1988), Durkheim (1893/2014), or Wright (2015). For that matter, Weber's multifaceted interpretation of class, while commonly overvalued and devoid of a relational aspect that makes Marx's treatment so illuminating, is much more useful than offhanded, commonplace renderings (Roth & Wittich, 1978).

We don't just see through class structure using the concepts of these extraordinary social theorists. Instead, we see class structure and its consequences for the first time. It may be that Berger sometimes confuses himself by giving undue attention to sociology's estimable debunking power, and he forgets that sociological concepts enable us to see a socially structured world to which we otherwise would be blind (Berger & Kellner, 1981). This is not inconsistent with debunking, and the two processes may take the same form. Still it's easy to get confused when working with a discipline that both reveals the truth and uncovers falsehoods, that both discloses structure and intelligently debunks conventional wisdom.

The strength of sociological determinism

Berger's sociological determinism is strengthened when he quotes the German sociologist Arnold Gehlen as holding that a social institution is "a regulatory agency channeling human actions in much the same way as instincts channel animal behavior" (Berger, 1963: 87). Berger develops this idea unambiguously when he affirms that "the social channeling activity is the essence of institutionalization, which is the foundation of the social construction of reality" (Berger & Luckmann, 1966: 181–182). Clearly this means that institutions pattern human behavior in predictable ways and behavior is compelled to take shape and direction so that it is socially acceptable.

It's easy to see that conduct directed in the manner described by Gehlen will have a taken-for-granted character. The various kinds of conduct that can occur, moreover, will be institutionally limited, rendered both predictable and controlled (Berger & Luckmann, 1966).

Still, Berger's judgment that we sometimes use the social roles we occupy to explain what otherwise would be egregiously unacceptable behavior is a useful insight (Berger, 1967). Again, however, he would do well to make clear that this applies primarily to roles that are products of offhanded secondary socialization rather than primary socialization or thorough deep secondary socialization. It would also be useful for him to lend credibility to his position by explicitly stating that established social institutions according to which we have been born and bred and live out our lives are not ephemeral fictions but are concretely and coercively real, as he emphatically explains in other places (Berger, 1963; Berger & Luckmann, 1966). This brilliantly insightful observation is sufficiently counterintuitive that it merits functional redundancy.

In the case of the public executioner, an example used by Berger, the hypocrisy and bad faith inherent in the claim that he is merely behaving as his social role requires seem thoroughly unambiguous. A killer is a killer. Nevertheless proponents of capital punishment might not see it this way. They might, moreover, point to the physician who performs abortions because he is a physician. For those who share the view that abortion is an abomination, the same charge might apply: A killer is a killer (Berger, 2011).

Think of Inspector Javert in Victor Hugo's, *Les Misérables* (Hugo, 1862/2013). The rigid determinism fostered by his unlikely life suffused him with one and only one world view, and one and only one social role within that world view, so fixed and inflexible that others' departures from his expectations were simply unfathomable. As a result, when faced with disruptions in what Javert took to be the only possible world, he found the departures from things as his socialization taught him they had to be intolerably jarring, and he committed suicide.

Javert was born in the prison. One imagines that both his primary socialization and virtually every aspect of his secondary socialization were prison-bound. His generalized other was a product of prison life, and his psychological makeup provided no avenues of escape from the harsh and inflexible dictates of the prison. Javert's generalized other was the result of an unusually deterministic and regimented set of circumstances that seemed certain to eventually have tragic consequences.

Images such as Javert suggest to me that Berger sometimes pushes much too far the notion that, when all is said and done, we are free of the social constraints built into sociology's deterministic conceptual apparatus (Berger, 1963). It's as if he's trying to make amends for some sort of unduly deterministic sociological misadventure by treating sociology as a set of excuses for doing the wrong things and thereby harming others with impunity.

He's telling us, in effect, that we can step out of any social role whenever we like. This means, of course, that we are fully aware of all the roles we play, an idea sharply at odds with the notion that roles into which we have been thoroughly socialized have an unself-conscious nature, and that we can't simply get rid of them. Berger's example of an executioner is striking, but it also refers to a role that's prescribed rather than inscribed (Berger & Luckmann, 1966).

I'm using the locution "inscribed" to again express the view that one of the differences between roles is the ease or difficulty with which we can ignore them. Some roles really are the sort that we can step in and out of as the mood strikes us. Other roles, by contrast, the inscribed roles, are ignored only with great difficulty and even pain. This difference, I'm sure, corresponds to the determinative strength of the institution the role represents, as well as the well-known distinction between primary and secondary socialization. Even if he restricts himself to roles acquired through secondary socialization, Berger would do well to examine variations in institutions' holding power, and perhaps explain why some roles are more effectively internalized than others, and some institutions are more forcibly directive than others.

For the most part, Berger is not convincing when he separates his sociology from determinism. Comparing the quality of the first five chapters of *Invitation to Sociology* (Berger, 1963) with the last three, I can only conclude that it's much easier to make the case for strict social determinism than for free will, especially when writing in terms of a discipline for which a substantial degree of social determinism is essential to justify its existence (Berger & Kellner, 1981; Berger, 1986; Berger, 1967). This is something that Berger, to his credit, subsequently acknowledges. Perhaps that helps explain why the following quotation is so persuasive:

> Institutions as historical and objective facticities, confront the individual as undeniable facts. The institutions are there, external to [us], persistent in their reality, whether [we] like it or not. [We] cannot wish them away. They resist [our] attempts to change or evade them. They have coercive power over [us], both in themselves, by the sheer force of their facticity, and through the control mechanisms that are usually attached to the most powerful of them (Berger & Luckmann, 1966: 60).

In my view, rather than being merely convenient placeholders, established sociological concepts are real. The fact that they are created, often unselfconsciously, by human beings, in no way diminishes the truth of that observation. Given the immediately foregoing discussion, how could it be otherwise?

Berger's simple but compelling observation that his primary interest is in the knowledge that guides everyday life (Berger & Luckmann, 1966: 19) is, in my view, a tacit endorsement of determinism, and it reinforces the reality of sociological concepts. After all, he's not saying that we willfully and knowingly conjure up and manipulate these ideas. They are simply there and they guide routine behavior. The case for determinism is strengthened as Berger continues:

> I apprehend the reality of everyday life as an ordered reality. Its phenomena are prearranged in patterns that seem to be independent of my apprehension of them and that impose themselves upon the latter. The reality of everyday life appears already objectified, that is, constituted by an order of objects that have been designated as objects before my appearance on the scene (Berger & Luckmann, 1966: 21–22).

The reality of everyday life is organized in a way that seems routine and unexceptional. Some things are close by, within reach of my hand, while others are remote and beyond the range of my senses, but I know them from experience of various kinds nonetheless. I know them perhaps only because I have read about them in books and magazines, but I still know what is real and what is not, and what may be, at least in principle, accessible (Berger, 1963).

In addition, my everyday world of commonsense is an intersubjective world (Berger, 1963). I share my commonsense understanding of the everyday world with others, especially friends and coworkers. Both they and I take this shared world of reality for granted, and we use it to unreflectively communicate, especially in conventional conversations whereby each of us unwittingly but effectively affirms the taken-for-granted reality of everyday life and our place in it for the other (Berger & Luckmann, 1966).

The immediately foregoing is the sort of statement that lays claim to the value of friendship in a socially compelling way. Assertions of this kind are fairly common in Berger's work, and though I know that they serve a sociologist's utilitarian purpose, they strike me as charming, giving the lie to gossipy, snarky, acerbic renderings that occur so frequently in, as an odd example, the novels of John Updike, like *Rabbit Redux* (Updike, 1971). Perhaps Berger's life has been more pleasant than Updike's was.

Conceptual elusiveness of free will

That said, it's very much to Berger's credit that in an explicit and unequivocal way that took me by surprise, he acknowledged that "Freedom cannot be disclosed by the methods of any empirical science, sociology most emphatically included" (Berger & Kellner, 1981: 96). Berger's case for free will resides in part on his insistence that the substance of sociology and other empirical disciplines is always "partial," leaving open the possibility of people behaving freely. Beyond that, the following quotation is especially illuminating:

> The relevance structure of any empirical science is limited to perceptions of causal determination ... for sociology specifically there is no way of dealing with human reality except within the framework of systems of social determination (Berger & Kellner, 1981: 96 & 98).

This is an extremely interesting observation, one that would never have occurred to me. As I read it, it affirms social determinism in an especially convincing way. After all, Berger is referring to what sociologists can discern, not to things as they actually are. Berger's insights into what we can expect of sociology are extremely interesting and the product of a keenly insightful sociological consciousness.

Invitation to sociology and determinism again

We have already noted too many times that the first five chapters of *Invitation to Sociology* (1963) are an extremely compelling argument for social determinism. It's paradoxical, though not necessarily dishonest, that the author of those chapters should then take the position that, when all is said and done, we are free to do whatever we want. But if that is the case, why study sociology?

When Berger asserts that "deception and self-deception are at the very heart of social reality" I want to ask him just why he bothered to write the first five chapters of *Invitation to Sociology* (1963). I learned long ago that explanatory science is impossible without determinism. Berger's misguided efforts to debunk all of sociology in the concluding chapters of *Invitation to Sociology* (Berger, 1963) stand in stark contrast to the truly fine first chapters of the same book.

Nevertheless, the flawed and incomplete nature of any empirical discipline seems to be emphatically true. If our behavior were in fact socially determined, completely and in every respect, sociology would not be able to discern this. Again, I laud Berger for acknowledging the need for some sort of explanation for his insistence on free will, and for making the best argument he had. Even though it's very brief, Berger's explanation is the most credible one I've seen by a sociologist.

19 Modernization augmented

Given the variability among people and groups it is obvious that socially determined interests will also be different, and often not compatible. As with the old middle class and the new middle class, group differences will give rise to a varied set of ideologies each embodying the specific interests of a distinctive group. The ideologies may be true, false, or ambiguous, but use of the term ideology need not invoke evil or disreputable means and ends. Insofar as groups have different interests and hope to lay claim to the same prestige, political resources, or material means of pursuing distinctive ends, ideologies will be one of the social tools they use to advance their interests.

However, this may be, in modernized societies the primacy of technology-intensive production in determining the character of a social system that will still be in place. Whether modernization takes the form of capitalist or socialist organization the need for massive political bureaucracies will be retained. As modernization makes its way into less developed territory, moreover, we can expect the consciousness of the indigenous people to initially clash with the modernizers. An essential prerequisite of modernization is to diffuse the conflict. Insofar as the rational ethos of the modernizers replaces the traditional beliefs and ways of reasoning of indigenous people, Berger's concept of a "Homeless Mind" will again manifest itself to the discomfort of those imbued with a traditional mind-set.

Berger, briefly augmenting his earlier definition of modernization, subsequently defines it as a social process that results in the transmission of specific collections of institutions and corresponding contents of consciousness (Berger, Berger & Kellner, 1973). This sounds very much like the definition of modernity I contrived at the outset.

Berger then makes what I think is a very unfortunate error in judgment, one that bespeaks thoroughgoing methodological naivete rather than theoretical sophistication. It also seems pointless. Berger holds that since, in *The Homeless Mind* (Berger, Berger & Kellner, 1973), he has:

> emphasized that modern consciousness is an assemblage or constellation of specific themes … this means that although [he] has not introduced

> measures as scales, [his] approach has been quantitative rather than typological (Berger, Berger & Kellner, 1973: 119).

He goes on by observing that because he has tried to analyze the content and style of modern consciousness with reference to a variety of cognitive and normative themes, the "frequency and intensity" with which these themes occur provide a quantitative accounting of the prevailing degree of modernity (Berger, Berger & Kellner, 1973: 120). Unfortunately, Berger does not tell us how to gauge the intensity of themes, but, in this instance, that's the least of his sins.

Why Berger tries to make the case that his work on modernized consciousness is quantitative is another mystery. All along, I thought that he was relying primarily on his usual method of description and interpretation, namely "sociological tourism," which would have been a good way to proceed.

Certainly, if Berger were going to make the assessment he has just offered, it should have come much earlier so we knew what he was doing and could see how his approach developed. As things stand, his claims as to having produced a document that, at least in part, is based on quantitative research falls somewhere between silly and preposterous.

Variability of modernization and of modern roles

Berger tells us that modernization is a worldwide phenomenon. It has manifested itself, however, in enormously varying degrees, and is least inclusive and effectual in the so-called Third World, where some nations contain merely pockets or parcels where modernization can be found. Given this high level of variability, one can see why Berger would want to measure and quantify modernization, but his approach to this difficult task is never explained or even illustrated. Berger never gives us so much as a single example of the measurement of modernization in any of his work (also see Horkheimer, 1947/2013).

Suffice it to say that there may be places where prevailing social circumstances are so different from the processes and substance of modernization that indigenous people may not recognize modernization as real. Elsewhere manifestations of modernization may be construed as novelties, similar in various functional and non-functional ways to that which already exists. In still other locations instances of modernization may be taken to be improvements on established ways of accomplishing essential ends with better tools, machinery, and procedures. Clearly, the world is fraught with multiple realities, each depending on the social circumstance and vantage point of those who apprehend it (Berger & Luckmann, 1966). However, while acknowledging these ordered differences may vaguely suggest an approach to measurement, perhaps intimating categories on an ordinal scale, we're still far removed from actual measurement of any sort or at any level.

I have always found it especially interesting to consider the notion of multiple realities in light of the fact that it may pertain to people who have lived their entire lives in the same complex societies. A municipal trash collector and a privately employed electrical engineer are cases in point, selected merely because they present us with a sharp contrast. By virtue of the work they do, the socially constructed road they took to gain employment, differing levels of compensation and familial and other consequences, they live in different worlds. This helps us understand why Durkheim (1897/1997) was so interested in the division of labor in society.

With this example we are in no way passing judgment on the comparative value of either occupation, though I will acknowledge that I am grateful that the trash is collected and that I find cell phones annoying. Nor are we offering an evaluation of the kind of life, whether for a bachelor or a man with a family, either kind of employment can provide. These are simply sharply contrasting examples meant to clarify what we mean by the claim that two people born and raised in the same society can have very different perspectives on what they take to be the real world.

As a sociologist of knowledge, Berger would have a compelling interest in both worlds and the contrast between the two. Identifying the sources of the contrasts in a way that enabled them to be transformed into concepts of general applicability would be especially valuable to the sociologist of knowledge as a social scientist. This could be an important part of the process whereby we determine the relationships between thought, knowledge, and social organization (Berger & Luckmann, 1966).

Insofar as members of the two very different occupational groups do their work with self-directed diligence, a strong sense of personal responsibility, and each invests his work with a suitable level of attention to productive detail, there are many who would wholeheartedly subscribe to the view that they must be members of a capitalist society (Berger, 1986). Berger might summarize this set of circumstances by invoking P12 and P13, as I have numbered Berger's propositions in *The Capitalist Revolution* (1986):

> P12: Given the social and cultural bases of Western civilization, capitalism is the necessary but not sufficient condition for the continuing reality of individual autonomy.
>
> P13: Certain components of Western bourgeois culture, notably those of activism, rational innovativeness, and self-discipline, are prerequisites of capitalist developments anywhere.

Anticipating ideologies

For reasons that may now be familiar, a set of circumstances such as that just described, may very well give rise to ideologies, ways of thinking that benefit one group perhaps at the expense of the other. Trash collectors may think of themselves as underpaid, overworked, and undervalued when compared to the deskbound, handsomely compensated work of the electrical engineer. Electrical engineers may see themselves as modern men of the most usefully productive kind whose status should be acknowledged as unsurpassed by any others.

The truth of these perspectives is immaterial. The point to be made is that there are very few – very likely no – activities to which ideologies, ways of thinking that give priority to particular activities and modes of being, are not attached. As we have seen, however, self-interested ideologies need not be evidence of undue selfishness, overweening pride, or pure evil, but they certainly seem to be ever present.

In the United States and Western Europe, modernization has taken place largely as a consequence of internal developments. In the Third World, by sharp contrast, modernization occurs largely as a set of institutions and styles of consciousness imposed or imported from the West. Little wonder that those who are not comfortable with modernization condemn it as a process of imperialistic Westernization. Ideological preconceptions notwithstanding, in a very real sense, it seems to be just that.

Berger's antipathy toward what he has termed Third World rhetoric or groundless ideology, as in "imperialistic Westernization," might have been tempered if he had access to a recent book by Max Friedman titled *Rethinking Anti-Americanism* (Friedman, 2012). Friedman's view may be summarized as follows: International actors sometimes characterized as inexplicably anti-American and anti-modernization actually change their positions as America's policies change. And the same holds for the position of the United States with regard to Russia, China, and the broad span of Middle Eastern nations: If their policies are consistent with ours, we can work together, or at least not be at each other's' throats (Friedman, 2012). Friedman's empirically grounded point is that other nations, including those in the so-called Third World, respond to The United States' intentions and actions rather than the simple fact that it exists.

Given the piecemeal nature that often characterizes Third World modernization, moreover, its nature is often grossly distorted and a source of considerable social, cultural, and physical pain. Technology-intensive production and bureaucratically organized rationality are still essential, as is the case in the more thoroughly modernized world. However, the importance of other, namely secondary modernizing agencies such as urbanization and education may be enhanced. Any quantitative analysis of the connection between technology-intensive development and styles of consciousness would have to measure and incorporate these confounding factors. Thus, the level of difficulty of the quantitative analysis Berger unexpectedly proposed continues to escalate.

A final thought: P12 and P13 may have substantial, even though untestable, merit, but they also have the sound of the romanticized individualism we mentioned above. Does either or both ring true based on one's experience, or are they merely idealizations that have the character of applause from true believers who, in much the way as Berger, have discovered capitalism?

Primacy of technology-intensive production

Nevertheless, Berger hypothesizes that the degree of development of technology-intensive organizations of production is as important in the Third World as it is in advanced, intensively modernized societies. According to Berger, "the technological transformation of material life is at the core – indeed *is* the core – of what is commonly meant by modernization" (Berger, 1986: 35). Berger puts this in strikingly direct terms as follows: "Put in its simplest form, this would mean that there is a direct relationship between GNP and consciousness" (Berger, Berger & Kellner, 1973: 122). Quantifying a variable as complex and poorly understood as "consciousness," by which I assume Berger means "style of consciousness," would be a difficult and uncertain measurement task, though measurement efforts as ambitious as this are undertaken every day (Neuendorf, 2001).

The previous paragraph seems to posit the existence of a straightforward linear relationship where, given the complex and multifaceted nature of style of consciousness, such a readily understandable connection would be difficult to discern. Of course, consciousness could be measured and expressed as a single variable, perhaps using principle components analysis or one or another variant of factor analysis, but Berger includes so much more than the effect of GNP into consciousness, or style of consciousness, that even with good diagnostic statistics for the analyses just suggested, I would be suspicious (Walkey & Welch, 2010).

However, even though Berger's strong statement is inherently ambiguous, we will take it to mean something pertinent and, in simplified form, interpretable: As the technology-intensive economy grows, more people become imbued with a distinctive style of consciousness, whatever form it takes and however complex it is, it resembles that which is common in more modernized nations. As a result, characteristics such as mechanisticity and componentiality are more frequently in evidence. And so are alienation, anomie, and egoism.

For a sociologist who is usually careful to distance himself from mono-causal interpretations, Berger's position on this issue is unexpected (Berger, 1986). He is clearly making the case that a technology-intensive economy is an extraordinarily powerful agency in fostering promotion of constituents of the style of consciousness that is most closely related to modernization. Bureaucracy, the other primary agency, and all secondary agencies are important but have fallen behind.

Hypothetical process of modernization

Berger then qualifies his position by acknowledging that any piecemeal process of modernization may initially require only a relatively small number of people who are imbued with modernized cognitive content and a modernized style of consciousness. Insofar as still others are involved, they will fill the most menial positions, doing simple though often physically demanding and dangerous work that requires nothing in the way of a modernized point of view (Berger, Berger & Kellner, 1973).

These unfortunate actors may find themselves physically, socially, culturally, and psychologically dislocated, suffering the anguish caused by alienation, egoism and anomie. They may be torn away from their traditional patterns of living, while never benefiting from any of the positive aspects of modernization or a modernized style of consciousness. Deracinated and relocated in a thoroughly alien environment, modernization may forever remain as foreign to them as when they first left their homes.

They may know little if anything of the well-favored world of the privileged few, and their common stock of knowledge will be theirs alone. Were they, for whatever reason, admitted to participation as full-fledged citizens in the world of their superiors, they would almost certainly lack the taken-for-granted or recipe knowledge that they would need to get by from day to day.

People in circumstances as devalued as those just described serve emerging institutions of modernization without ever becoming members or benefiting in any but the most meager way. To their great misfortune, they have been relegated to the status of not belonging anywhere. This is homelessness in its most extreme and painful form, and if it's understood as such deserves the status of a crime against humanity (Robertson, 2006), though, as Berger sees things, this is not a value judgment for a sociologist to make.

Presumably, rudimentary institutions would form among such devalued groups. They would not be the sort, however, that entail the specific aspects of modernization. Instead, they would be emergent institutional forms of little external clout, perhaps initially comprised of unrecognizable institutional detritus from the structures of modernization that they serve, much as a slave serves his master.

I think, however, that if this rudimentary social system took on the characteristics of a nascent capitalist society, it would have to acknowledge P14, especially as it became more complex:

> P14 Capitalism requires institutions that balance the anonymous aspects of individual autonomy with communal solidarity; among these institutions are the family and religion (Berger, 1986: 113).

This simply means that anomie and egoism, concepts borrowed from Durkheim and meaning cultural deregulation and absence of a sense of belonging would have to be subject to institutional regulation to avoid various kinds of social pathology.

If this sort of social circumstance were to continue across generations, it is likely that a separate, even if devalued, social system would form. The dialectical process of externalization, objectivation, internalization would inevitably be at work in the formation of the institutional components of this new system.

Perhaps an internal division of labor would emerge, with some acting as hunters, some as fishermen, others as farmers, and still others in numerous emergent categories. Membership in one occupational group or another could be passed from one generation to another, using both observation and linguistic instruction to socialize the next generation with the requisite knowledge.

In time, knowledge would be distributed in a way that meant that not all knowledge was part of a common stock. Instead, specialized argots would emerge, each organized around a different occupation. At this juncture, the institutions needed to address problems attendant to anomie and egoism would become increasingly important,

Stratification in the form of classes, given the nature of our world, would be sure to manifest itself to the advantage of some and the disadvantage of others. If a money economy had taken shape, the distribution of wealth and income in a way that emulated dollar terms would provide a convenient exchange basis for the class system. Monetary links with the occupational distribution are likely, as is private control of land in a place where land was once a purely public resource.

Berger makes an insightful and interesting observation regarding economic activities such as those just mentioned. Assuming that the hunters, fishermen, and farmers are not already familiar with the requirements of these roles, in time commonsense sciences of hunting, fishing, and farming may emerge. By science, in this context, Berger means an observable body of routinized practices that more or less predictably yield the desired economic results. Insofar as these habituated activities are passed from person to person and generation to generation, linguistically and through shared experience, we are witnessing the emergence of institutions of hunting, fishing, and farming (Berger & Luckmann, 1966).

In effect, we are seeing the development of a new and separate, even if subordinate, culture and social system. As with any society, its common stock of knowledge would be manifest in and shared through language. Thus, even those who have not experienced a very substantial part of the experiential world that may be directly known to others can share in the common stock of knowledge through the medium of language. Given the passage of generations, language may be the sole repository of transmitted information that participants come to take for granted.

The shared language available to all for the incarnation of their experiences is a medium whose importance is impossible to exaggerate. This applies to the maintenance of established societies as well as to the creation, and subsequent maintenance, of new ones. Human beings would not be human without language, and their social systems would not exist even in the imagination. After all, without language imagination would not exist (Mead, 1934/1967).

Those benefiting from the labor of this subordinate society would thus have access to a supply of menial labor for the long term. Perhaps an overabundance of non-modernized labor would form, with neo-Malthusian consequences that would be impossible to foretell. This is speculation, of course, but the world is a precarious and unpredictable place.

Ironically, however, the more thoroughly institutionalized this ancillary social system became, the more predictable and tractable its inhabitants would likely be. After all, we tend to become, and to understand ourselves in terms of, what we do. Those born and raised in devalued circumstances are not exceptions. In due course, they understand themselves in the devalued terms in which they live. If these were the only terms they had acquired, their ability to adopt an evaluative perspective on their circumstances would be very narrowly constrained (Berger & Luckmann, 1966).

As each generation succeeded another, the reality of the circumstances of members of the devalued society would become more compelling. Their origin as something else – anything else – would be lost across the generations, and the collection of institutions that constitutes their society would be the only social world they knew.

Surely they would be aware that there was a society to which they were subordinate. Insofar as they completely lacked knowledge of everyday life within it, however, it would remain alien and opaque. While human beings are perfectly capable of understanding multiple realities, doing so is premised on access to information, even if it's fictitious or preternatural, with which such an alternative social reality can be construed (Berger & Luckmann, 1966).

We have taken a roundabout route, but I think we can now begin to imagine the uncertainty and harshness that might be likely to accompany modernization if it were limited to the imposition of technology-intensive organizations of production. Technology-intensive production is a powerful agent of modernization, but it is also painfully incomplete. Whether or not an ancillary social system such as that described above has ever been approximated, we see that it very well could be, a peculiar product of modernization, indeed.

Nevertheless, Berger continues to focus on development of the technology-intensive economy. He takes the position that every step in the modernization of a technology-intensive economy prompts a corresponding change in consciousness. The economy is, indeed, a powerful agency of modernization.

This compelling claim offered by Berger, however, is one that is easy to understand but hardly typical. Qualitative transformations are only one aspect of the ongoing, often gradual and not conspicuously discernible, process of modernization.

The role of political bureaucracies

In spite of the position of priority Berger has assigned to a technology-intensive economy in a modernizing country, the other primary agency of modernity, rational bureaucratic organization, also plays a significant role in promoting modernization. This is especially true in countries where the state adopts the role of a mobilizing agent for modernization, and the state commands the allegiance of the population in the modernizing process (Berger, Berger & Kellner, 1973).

South Korea is a primary example of a nation that modernized in good part through the consistent force of the political will of successive, like-minded administrations, buttressed by the military. In countries such as this, the most highly valued occupational opportunities are often found in government – in bureaucratic organizations – given that government wields enormous power in what is otherwise a hard-nosed capitalist country (Brazinsky, 2009).

Berger, however, takes a different view of modernization in South Korea. Though he uses only an anecdote or two to make his case, the orderliness of a group of South Korean school boys in an uncomfortable and tedious set of circumstances, for example, prompts Berger to ask, "Were we watching the Korean economic miracle in its biographically primordial roots?" (Berger, 2011: 147). Perhaps he was, but it's this sort of offhandedness that may prompt readers to wish that Berger were less inclined to sociological tourism, especially when it takes episodic form, and that he were more interested in some sort of more or less rigorous empiricism. I think that Berger has come to take the Protestant Ethic and rough approximations to same entirely too seriously and too broadly construed, and maybe he sometimes just doesn't bother to think about it.

We have seen that sociological tourism can be labor intensive and qualitatively rigorous. However, the notion that "Only a wealthy society can afford misbehaving children" is clever and witty, but in addition to being of doubtful veracity, it's not the sort of thing that effectively substantiates Berger's claim that he has found the Protestant Ethic alive and well in East Asia. Sociological tourism can be carelessly anecdotal, too.

Third World bureaucracies

It is important to acknowledge that people from modernized nations often find Third World bureaucracies cumbersome, slow, inefficient, and more of an obstacle to modernization than a facilitator. One reason for this is that many aspirants to positions in a governmental bureaucracy, intelligent and hard-working though they may be, fail to understand that fostering the effective functioning of a bureaucracy requires training, skill, and experience.

That this observation is hardly new follows from the fact that it was known to Weber writing in 1922:

> The question is always who controls the existing bureaucratic machinery. And such control is possible only to a very limited degree to persons who are not technical specialists. Generally speaking, the highest ranking career official is more likely to get his way in the long run than his nominal superior, the cabinet minister, who is not always a specialist (Weber, 1922/1978: 224).

In time, recognition of the requirements of bureaucracy for well-trained and suitably socialized participants, even in the Third World, can prompt the emergence of a new middle class, one consisting largely of well-paid, upwardly mobile employees of the state. Understandably, this group's appreciation of the nature of modernization will be more closely allied to the style of consciousness intrinsic to a political bureaucracy than that of technology-intensive production. As we have seen, however, the two styles of consciousness are similar.

The primacy of political bureaucracy as a means of promoting modernization is something we have not seen before and that seems inconsistent with Berger's earlier emphasis on development of technology-intensive production as not only a primary agency of modernization, but *the* primary agency that emphatically is of first importance. Nevertheless, Berger clearly acknowledges that effective employment of state bureaucracies may be a prime mover in a process of modernization.

Modernization and disruption through capitalist and socialist endeavors

In the early 1970s Berger, acknowledged that modernization through means of the capitalist mode of production was often painfully disruptive – even massively destructive – of traditional forms of life, organization, religion, custom, and governance. This experience had a radicalizing effect on many. Berger claims, however, that socialist endeavors at modernization were at least as destructive, but they generated less hostility because they were ostensibly geared toward attainment of more or less well-defined goals that seemed humane and otherwise more attractive than the ongoing process of capital accumulation (Berger, Berger & Kellner, 1973).

Berger provides us with an interesting observation that helps us to understand the appeal of socialism in The Third World, namely if modernization can be described as a spreading condition of homelessness, then socialism can be understood as the promise of a new home. Berger, however, sees in socialism one more force toward centralization of power, and therefore a possible precursor of emergence of a totalitarian state. It is useful to recall, however, that Berger, throughout his later work, never construed totalitarianism to be a barrier to modernization (Berger, 1986). Berger concluded, in effect, that while capitalism was necessary to sustain democracy, democracy was not a prerequisite for the emergence and development of capitalism.

It is extremely interesting that in 1973 Berger had not yet found reason to become a determined proponent of capitalism, a turn in his intellectual development that came later with his experiences in East Asia. *The Homeless Mind* (Berger, Berger & Kellner, 1973) includes fairly numerous references to Africa. Evidently Weber's Protestant Ethic had not taken root there, though Berger admonishes his readers that hard work is something that African's embrace as readily as others.

Collisions of consciousness

As a phenomenon experienced by members of the least developed Third World countries, Berger attributes an almost a supernatural, miraculous character to modernity, however adequately or inadequately we define it as its possibilities are first fathomed. After all, modernity is connected with the expectation, and occasionally the reality, of an escape from hunger, disease, and early death, unpleasant and unforgiving processes that were long taken to be inescapable facts of life in much of the Third World.

According to Berger, the miracles of modernity may become well known, at least from a distance, to Third World inhabitants even if they are not substantiated by promised or expected changes in their circumstances. Though the triumph of the introduction of the rudiments of mass communication and commonplace hearsay may make Third World inhabitants seem naïve and hopelessly gullible, a very misleading set of circumstances, the mystique of modernity is difficult to ignore, and it is something that affects most of us in much the same way.

Perhaps in the Third World, this is due to the fact that, as time passes, the sources of what are taken to be the near-mystical and compelling messages of modernity are sanctioned by a proliferation of agencies. These include even the most elementary levels of schooling, as well as informal reports offered by those who have had experiences outside the cloistered informational self-sufficiency of the isolated rural village.

The message and promise of modernity may be especially plausible when its source is governmental agencies' stipulation that they will accomplish realistic-sounding reforms. When the promises are not kept, since they seemed genuinely possible and likely to be accomplished, it's not surprising that a rebellious, conflictual response may eventually be in the offing.

For those with the closest ties to traditional symbolism, beliefs, and practices, some sort of collision with the encroaching ethos of modernity is virtually certain. For many, however, the allure of modernity is sufficiently strong to prevent this. The latter group is often made up of those who hope to move to large urban areas.

This holds even though cities often offer nothing but a more concentrated form of poverty. But urbanization, too, according to Berger, is infused with a seductive mystique, one that is manifest in its sometimes frenetic level of activity, something that migrants from the static environment of a Third World rural village hope eventually, in one way or another, will bring improvements to their lives (Berger, 2011).

Cities are suffused with symbols of modernity, attractive because of what they seem to promise. According to Berger (Berger, 1986), Third World governments are often compelled to manipulate hollow symbols of modernity simply to maintain stability and keep the population at large tractable, willing to wait for whatever it is that modernity will bring.

This sounds very much like the irrational stupidity that has traditionally been ascribed to people in the Third World as well as impoverished places in the modernized world such as the Appalachian United States (Bickel, Tomasek & Eagle, 2000). Berger does not trouble himself with efforts to affirm or discount the validity of this interpretation. Berger's certainty in making judgments such as this sometimes spills over into a sort of matter-of-fact arrogance, additional instances of his sometimes-undisciplined approach to research.

However, the knowledge that agencies of modernity bring to the Third World is not invariably specious and manipulative. Through schooling and a variety of other means, the content and organization of knowledge, along with the style of consciousness, are transformed in ways that are sharply at odds with tradition and may be useful.

Schools, especially, serve to legitimate the new knowledge and the social consequences of its distribution. Still, for anyone who has visited ill-equipped, underfunded schools in the United States, where well-intentioned teachers are obliged to teach subjects for which they have no training, it comes as no surprise that the value of what is taught is sometimes a good deal less consequential than one might imagine (Bickel, Tomasek & Eagle, 2000). Instead, the school as a quasi-sacred institution bespeaking modernity endows those who attend it with a favored status that may actually betoken nothing more than that they have participated in institutionalized rituals imposed or imported from the modernized world.

Many forms of rationality

Modernization, as Berger effectively explains it, brings rationality in many forms. Science-based technology itself is unthinkable without the unquestioned ascendance of functional rationality over myth, fable, poetic imagery, and other ways of making sense of the world that predate, and to some extent co-exist with, rationality (cf. Santayana, 1967). Rationality is also inherent in modern bureaucratic organization with its category for everything and everything with a rationally justifiable category. In addition, capitalism, whatever its virtues and abuses, is also thoroughly infused with an ethos of rationality. Marx himself understood this, making the case that capitalists were caught up in the rationally calculable pursuit of profit to the exclusion of all else (Marx, 1844/1988). These are instances of modernization in the form of functional rationality that we have already discussed.

The encroachment of modernity brings with it the kind of instrumental rationality that has made modernization so consequential in the heavily industrialized and otherwise developed world, and the transformation in Third World styles of consciousness has been overwhelming. A world that was once accepted, for better or worse, as home for traditional residents of the Third World becomes something quite different. Instead of a home to be accepted largely as it is, the world is now a place to be transformed and purposefully exploited.

Sometimes the transformation is made to secure outcomes that inhabitants of already modernized countries would judge to be wholesome, even essential to living a decent life. The eradication of disease, reductions in infant and maternal mortality rates, and alleviation of lethal poverty are obvious examples.

However, not all outcomes of rational activity as the modernized world understands it are so readily defensible for one and all. Much of the Third World is rich in raw materials that bring substantial profits to fortune-hunting individuals and profit-seeking corporations (Todaro, 1985). It seems unlikely that many members of intensively modernized social systems would condemn use of Third World countries as repositories of oil, copper, tin, tungsten, and other sources of profit, but procuring them need not be essential to the lives of inhabitants of the Third World. In fact, it may be just another instance of shameless exploitation, though this is my speculation, not Berger's. Berger, as we have learned, is much more likely to see it as the work of enterprising organizations and individuals (Berger, 1986).

Perhaps most important, instrumental rationality and its imposition represent a wholesale transformation in the world view of those affected. After all, is there anything really irrational about viewing the world as a home that merits our respect and that is not ours to exploit? In fact, this would be an example of what Weber termed Wertrationality as opposed to Zweckrationality. Inhabitants of intensively modernized societies, unless they are stupid or nakedly greedy, have learned the lesson that unbridled employment of instrumental rationality can eventually become even more threatening than disease and hunger (Horkheimer, 1947/2013).

Traditional modes of reality in the Third World

Berger gets an opportunity to use his neologism componentiality (Berger, Berger & Kellner, 1973) when contrasting the rationality of modernity with a conception of reality that apparently is common in the Third World. In some places reality is traditionally construed as having a living and interconnected bond for all of nature, with nothing separate or isolated. Individualism, among many phenomena peculiar to the modernized world, is inconsistent with this view.

As a result, when modernized styles of consciousness encroach, including, conspicuously, componentiality, a dramatic transformation in the way the world is apprehended is inevitable. A stark alternative to the everyday social reality with its objectively cohesive facticity that was once taken for granted as real is imposed.

Yes, sometimes we can purposefully distort the reality of everyday life to make that which appears inconsistent fit, made readily expressible in the commonsense language we are accustomed to using. However, when faced, as in the present example, with an alternative that represents a qualitative and complete transformation, our efforts at restructuring and repair quickly break down. Little wonder that a sense of homelessness becomes pervasive. This is egoistic deracination at its nakedly destructive worst.

Much the same is true of transformations in ways of calculating time. Observers who have noted what is taken to be the painfully slow functioning of Third World bureaucracies, may fail to understand that time may be reckoned in very different ways in the Third World.

We discussed this issue earlier with regard to the need for training and experience in the making of competent bureaucrats (Berger, Berger & Kellner, 1973). It is also true, however, that the rationally calculable and numerically precise view of time that is intrinsic to modernity may be inconsistent with Third World reliance on natural rhythms of various kinds in reckoning time. Given that, what are we to make of the following quotation, something that by now has no doubt been read by millions of serious people?

> Temporality is an intrinsic property of consciousness. The stream of consciousness is always ordered temporally. It is possible to differentiate between different levels of this temporality as it is inter-subjectively available. Every individual is conscious of an inner flow of time, which in turn is founded on the physiological rhythms of the organism though it is not identical with this … Furthermore the temporal structure of everyday life confronts me as a facticity with which I must reckon. That is, with which I must try to synchronize my own projects (Berger & Luckmann, 1966: 26–27).

I think the best way to interpret this paragraph is to acknowledge that at the most fundamental level time is the same for everyone. Whatever the variety of ways of reckoning time from one place to another, they have a deep, even visceral, commonality. The fact that the world we consciously share sometimes presents itself as sharply at odds with this means that, for good and sufficient reasons, most often to foster instrumental rationality, the ways that we commonsensically understand time have been socially constructed in a different manner. In whatever ways they are constructed, we cannot ignore them.

Thus, what may be taken as intransigent inefficiency born of ignorance, laziness, indifference, or stupidity on the part of Third World bureaucrats, may be better understood by acknowledging that time, as a manifestation of instrumental rationality, is alien to some Third World settings. Perhaps this is reason enough to say that in some instances Third World countries have been bureaucratized prematurely. Whether or not that is the case, it's easy to see that a common understanding of time is essential to getting along harmoniously in the everyday world.

Indigenous ways of reckoning time, moreover, can have a rationality of their own, a product of observation and understanding of how their traditional world actually worked in the social setting it occupied. Functional rationality is the interloper, and its imposition is disruptive, indeed.

Homelessness in perspective

It's difficult to know if one consequence or another has had the greatest impact through the inevitably painful process of modernization in the Third World. However, as inhabitants of intensively modernized societies struggle to find value, permanence, and meaning in their lives, it may occur to them that the weakening of religion as a source of lasting faith, certainty, and social cohesion, unshaken by temporal events, is painful and destructive. The pluralization and secularization of the Third World has dramatically weakened the traditional spiritual sustenance and social significance provided by religion.

I don't know if it's better to live a long and physically healthy life or one that is infused with unassailable meaning, but when modernization segregates religion into the private sphere, the world is transformed. Some may think this is for the better, and perhaps it is. But it's my view that for many the process whereby pluralism ostensibly fosters secularism is genuinely tragic, a value judgment that a professional sociologist should not make.

Besides, ubiquitous social maladies born of modernization – alienation, anomie, and egoism – all things we've discussed before are commonly taken to impose a purely psychological burden of greater or lesser weight. However, when one's body is stressed by enduring factors such as these day in and day out, it's hard to imagine that the physical toll is not an invitation to physical pathology. Given the nature of the modernized world of work and the processes of modernization manifest in less developed countries, it seems perfectly understandable that maladies such as panic disorder, generalized anxiety disorder, alcoholism, drug addiction, hypertension, and heart disease would reach epidemic levels. To my knowledge, however, there is very little literature on this.

On the odd chance that modernization brings with it counseling, psychotherapy, medication and other therapeutic manifestations of modernity, these are in no sense liberating. Instead, they merely embed people more firmly in the modernizing world. The therapies available need not be liberating but may be usefully understood as methods of social control (Berger, 2014).

Modernization drains swamps, purifies drinking water, and brings substantial improvements in peoples' diets in the Third World. But modernization is accompanied by its own assortment of pathologies. The trade-off may be worthwhile, but it remains true that modernization imposes many costs, some of which Berger seems unaware.

I think, by the way, that Berger's account of secularization through moving religion or anything else not intrinsic to the technology-intensive, expansively bureaucratic domains into the realm of the private is a relatively weak way of explaining what is going on. The troublesome outcome is secularization, whether due to pluralism of something else, and saying that the sacred has been moved to the private sphere seems a bit too indirect or oblique to make this immediately evident and coercive.

The most compelling account of instinctive opposition to modernization and the cost of losing the battle that I've ever read is Joseph Conrad's novel *The Heart of Darkness*. It's difficult to imagine a reader of average sensibilities not being changed by Conrad's frighteningly fictional but terrifyingly informative account. *Heart of Darkness* is the strongest rendering of naturally motivated opposition to the homelessness that inevitably accompanies modernization that I've ever read (Conrad, 1899/2014).

20 Modernity and its discontents: a suitably ironic closing

Technology-intensive production, the most fundamental component of modernization and modernity, may be troublesome in many ways. It presupposes and builds on functional rationality. Rationality, by its very nature, is an organizing phenomenon that imposes tight controls on impulses. It is thoroughly constraining, permitting only that which fits into the purposeful processes it imposes, and rejecting anything that cannot be tightly woven into the patterns it allows (Adorno & Horkheimer, 1944/2007).

When discussing rationality, Berger rightly invokes the Freudian concept of repression, making spontaneous expression difficult if not impossible (Freud, 1909/1995). That rationality can be unpleasant and, in the extreme, intolerable is easy to see. After all, managing one's emotional life is difficult, requiring an unnatural sort of discipline. Perhaps this is another source of maladies of modernization.

Similarly, the anonymity intrinsic to modern life is often troubling. With regard to anonymity, Berger invokes Durkheim's concept of anomie or cultural deregulation. I think, however, that egoism or lack of belonging or membership better characterizes the possible consequences of anonymity (Durkheim, 1897/1997). When we live in a world where many social relationships are fraught with anonymity, such as the world of technology-intensive production and rational bureaucratization, we inevitably live in circumstances where we have a limited sense of membership, and a shared culture, should there be one, is extremely thin. Again, we are homeless.

I have mentioned before that Berger is inclined to use anomie in contexts where both anomie and egoism seem to belong. He also uses anomie in place of egoism, or so my reading of his work suggests. He no doubt has reasons of his own for doing so, but I don't know what they are, and I find that what I take to be his repeated misuse of anomie in place of egoism disconcerting.

However that may be, an anonymous world is not one that promises fulfillment and emotional richness. Instead, its sterility presents us with another reason for being dissatisfied with modernization and modernity.

The complexity of the modern world is something that most of us have come to take more or less for granted. However, dealing with complexity may be quite laborious, especially when its sources are ever-changing. I still don't like cell phones, and it annoys me to see people playing with them, as if they contained all that is entertaining or interesting in the entire world.

This is unpleasant, to be sure, and it becomes even more difficult when the degree or kind of complexity renders the world uninterpretable. When the world becomes alien to us, the concepts anomie and egoism are quite well suited to indicate our predicament, and we have good reason to object.

Reading over the foregoing makes clear that the modernized, technology-intensive, bureaucratized world is one where formality, impersonality, purposeful indifference, and pragmatism are ever-present. The world of modernity tends to be an abstract world, one where lasting concreteness in interpersonal relationships is denied by the very design of institutions and has become increasingly hard to find (Berger & Kellner, 1981).

Being alone is something that all of us value from time to time, but when it is ever-present and inescapable, it is also maddening. We all, by our very nature, need concrete contact and naturally complex patterns of human interaction. To the degree that we are deprived of these, we become more and more discontented and dissatisfied with our circumstances.

Mobility is a circumstance that, though it can be viewed in a wide variety of ways, has acquired a more or less favorable connotation. Perhaps this may have its origin in the post-World War II era when upward social mobility in the developed world of the United States was by far a more pervasive fact of life than today (Blau & Duncan, 1967; Schwarz, 1998).

Even upward mobility, however, is not an unmixed blessing. In fact any kind of mobility, whether social, geographical, occupational, or any other type has a downside. In much less modernized places where mobility is largely unknown, "what one is and where one is are *fixed*, and one cannot even imagine how one could be anything different" (Berger, 1963: 48). Losing this sense of certainty must be overwhelmingly disorienting.

When we are mobile, in whatever fashion and in whatever direction, we leave behind the familiar, exchanging it for the alien, for a set of circumstances in which we no longer feel at home. We have made a choice, one among what may be a daunting plurality of choices, and as a result we become imbued with a sense of homelessness.

Choices such as this, especially given the frequency with which they sometimes occur, are most common in modernized societies where modernity has already come to prevail. That they betoken economic opportunity is often the case. That they bring a sense of homelessness is usually undeniable. That they imbue us with the sense that all truth is socially relative, it just depends on your vantage point, may not be so readily apparent, until vantage points actually begin to change (Berger, 1963). The same idea is sometimes rendered as "all truth is socially relational," using relative and relational interchangeably, though I think the latter usage does not discount the possibility of religious faith as truth, while the former does.

Much the same is true of the weakening of religion as a social force that gave meaning to our lives, and culturally satisfying accounts for our misfortunes and eventual demise. With the decline of religion certainty is largely eliminated. If mobility costs us our earthly home, then pluralism, secularism, and the emergent ambivalence of religion cost us our metaphysical home. Before death everything is precarious, and after death there is just death.

The relativism that follows from acknowledging that we live in a socially constructed, pluralistic world has profound implications for a topic of enduring interest for Berger, namely the sociology of religion (Berger, 2011). With access to essential information, sociologists can explain the origin and creation of any religion. Methodologically and theoretically, the process will be little different from studying any other institution (Berger & Kellner, 1981). The socially constructed reality of all structures, practices, and attendant meanings in the world as we find it make this abundantly clear.

Furthermore, the sociologist who is sufficiently familiar with the meanings that give religion its force and promise of fulfillment can come to grips with the very nature of religious experience. This requires no special, much less supernatural, gifts inherent in the sociologist, powers not to be found in the methodological and theoretical wherewithal of sociology itself.

It does, however, require the ability to take the role of the other, much as George Herbert Mead explained in *Mind, Self, and Society* (1934/1967). Just as Weber sought to explain the nature and consequences of Calvinism, including its manifestations in the consciousness of its adherents, properly trained and experienced sociologists can do the same for other religions.

Relativism, fundamentalism, and a middle ground

As with morals, however, the validity of any religion raises questions outside the purview of sociology and its practitioners. Sociology can neither debunk nor affirm the truth of any religion, nor can it give an approving nod to atheism. Sociology is limited to studying the enormous variety of religious forms in their worldly manifestations, but at that point reins itself in and comes to a halt (Berger, 1967).

Berger makes an extremely interesting and, as far as I know brilliantly original distinction in his discussion of relativism and fundamentalism with regard to religion and anything else. Relativism, as in "let's agree to disagree," precludes the rejection or condemnation of anything, an impossibly flexible compromise that limits civil and scholarly discourse in ways that prevent useful and defensible conclusions. As a result, relativism prevents the formation of a moral consensus essential to the survival of any social system (Berger, 1992).

Fundamentalism, on the other hand, yields a hopelessly rigid commitment to one and only one perspective, religious or otherwise, with the permanence of interlocked conflict taking the place of any mode of accommodation. Fundamentalism militates against the kinds of compromises and concessions that are essential to democracy. In consequence, fundamentalism may divide society into conflicting social entities, unable to reach a mutually beneficial accommodation, becoming one society whose inhabitants are unbound by common interests (Berger [Ed.], 2009).

Berger then makes a very effective case for a middle ground, something that I didn't think possible in anything but a squishy soft, ill-thought through, speciously everyday-clever sort of way (Berger [Ed.], 2009).

Yes, individuals may be zealously religious or reject all forms of religious faith. More likely, however, they will unself-consciously gravitate to a middle position (Berger [Ed.], 2009). Whether or not they sensed that the extremes of relativism and fundamentalism were ultimately untenable is something we cannot know. If asked to justify their position it would very likely be clear that they have not given it much thought, and they may refer, even if obliquely, to the distinction we have just made, making a claim that they are avoiding extremes.

Furthermore judgments such as this occur outside the relevance structure or point of view provided by sociological practice, belonging to a set of experiences intrinsic to a part of everyday life. This, of course, is another instance of what Berger terms "dual citizenship" (Berger, 2011).

Insights of sociological tourism

Berger bases his alternative interpretation on the myriad observations of "sociological tourism," a truly informal though labor-intensive way of collecting and interpreting data, and there may be substantial merit to his position. After all, having to choose among a multitude of conflicting religious orientations, a process fraught with uncertainty, gives rise to the same sort of anomic and egoistic circumstances as that yielded by a thoroughly secular world, one in which faith is hard to come by.

If one were in a perversely cynical frame of mind, the observation that modernization and modernity, for all their promise, have not completely eliminated poverty, disease, and death might seem a truly damning judgment. Yes, we're asking a lot, but modernization and modernity have long promised a great deal. It seems only reasonable and fair to ask proponents of these ideas to be consistently expansive in what they provide.

In a real way, modernization and modernity, when all is said and done, have the character of a revolution that has both succeeded and failed. Modernization has swept across the globe transforming the lives of billions of people. However, the failure of modernization to meet the unrealistic goals that many had for it in purely material terms is extremely disappointing. Too many have been left behind, and all of us remain subject to unhappiness and disease, and we are certainly mortal. All that joined with the discontents of modernity discussed throughout this manuscript give us a muddled picture. Success or failure? Take your pick.

In one of his most interesting observations concerning the discontents of modernization and modernity, Berger reminds us that the private sphere provides us with a place to be whole human beings and to regard others as such.

Berger presents the private sphere as providing a domain where we can, for the most part, give expression to that which otherwise has to be suppressed, say at work or while engaged in a transaction with a bureaucracy.

However, Berger also provides us with the insightful observation that the private sphere is "under-institutionalized" (Berger, Berger & Kellner, 1973). By this he means that freedom devoid of organization and purpose is also devoid of meaning. What, in short, does one do to fill time when there is nothing to fill it with?

Before going any further with our discussion of the under-institutionalized nature of the private sphere, I think it useful to emphatically point out that a surfeit of leisure is no longer the burden it may have been as the post-World War II era ripened into the late 1960s and early 1970s. Instead, more people are working longer hours just to make ends meet, and, all too often, still failing to do so (Schwarz, 1998; Newman, 1999). Perhaps this is one reason why styles of consciousness intrinsic to technology-intensive organizations and political bureaucracies spill so readily into the private sphere. The private sphere has been institutionally compromised in ways that Berger could not have foreseen when writing his earlier, most influential work, including *Invitation to Sociology* (Berger, 1963), *The Social Construction of Reality* (Berger & Luckmann, 1966), and *The Sacred Canopy* (Berger, 1967).

Furthermore, families are much less likely to have a stay-at-home wife and mother. Instead both spouses are apt to hold full-time jobs. This demographic fact contributed substantially to maintaining post-World War II levels of affluence for several decades.

However, the economic benefits that accrued in the United States and elsewhere under the auspices of modernization and modernity have been undercut by the gross maldistribution of income and wealth that seems inevitable when the notion that a largely unregulated form of capitalism is the best for all is taken too seriously (Krugman, 2009; Piketty, 2014). If modernization and modernity have brought us disappointments, this is one that Berger did not anticipate. Instead, his presumptions were consistent with the mindset that prompted the creation of Ph.D.-granting Departments of Leisure Studies at major universities.

Even when Berger is right, as when he proclaims the virtues and importance of what he and many others have termed mediating structures, his commitment to capitalism is so strong that it prevents him from doing fair-minded comparative analyses and tends to diminish the persuasiveness of his presentation. When his work is done under the auspices of the conservative American Enterprise Institute, an organization with a mission statement that includes promoting capitalism, readers' interpretations may become even more uncertain (American Enterprise Institute, 2014).

Nevertheless, it is very much to Berger's credit, and an acknowledgment of his indebtedness not only to Weber but to Durkheim as well, that he discusses mediating structures in a sure-handed and interesting way that clearly captures their value. In fact, mediating structure is a term attributable to Berger himself and one of his co-authors, Richard Neuhaus (Berger, Neuhaus, & Novak, 1994). This again demonstrates that Berger's contribution of new concepts to the repertoire of social scientists need not be limited to those that are as opaque, vague, and clumsy as the ones we found in *The Homeless Mind* (Berger, Berger & Kellner, 1973).

Mediating structures, at their best, provide a sense of membership in an otherwise egoistic world. In addition, they buffer and facilitate the relationship between individuals and mega-structures such as large corporations and the state itself. Rather than being alone and at odds with a menacingly overarching environment, mediating structures such as religious organizations, local labor unions, volunteer fire departments, and other voluntary associations provide a sense of affiliation and common purpose, a sort of organized protection against being overwhelmed by anomie and egoism (Putnam, 2000).

Durkheim hoped that occupational organizations could be formed that would mediate between individuals located in occupations of all kinds and what Marxists sometimes refer to as Big Capital (Durkheim, 1893/2014). For the most part, this did not occur, but Durkheim (1893/2014) had envisioned occupational associations as offsetting the emergence of egoism and anomie in an otherwise intolerably large and alien environment. Durkheim, I'm sure, would have enthusiastically endorsed Berger's interest in promotion of the development of mediating structures that take an enormous variety of forms (Durkheim, 1893/2014).

Berger, however, goes on and obliquely posits the existence of a relationship between his work on mediating structures and awareness of unintended and unwelcome consequences of the malfunctioning of the American welfare state, insofar as one exists. Of particular importance is the controversial phenomenon commonly referred to as welfare dependency, or simple dependence on the state for the necessities of life, say through distribution of food stamps (Berger, Neuhaus & Novak, 1994).

As with most critics of initiatives by the state to alleviate widespread and potentially destructive difficulties such as devastating poverty, Berger does not offer effective alternatives to deal with failures of a developed capitalist economy to provide employment, a living wage, or other economic essentials. As best I can determine, it is Berger's assessment that mediating structures would play that ameliorating role. How this would come about, however, is by no means clear.

Berger's ambivalence with respect to a large and active central government may be something he first experienced long before he "discovered capitalism" (Berger, Neuhaus & Novak 1994). He takes the position that the welfare state has had the unintended consequence of weakening mediating structures and undercutting their emergence. It's not clear why Berger has adopted this position, but it's at odds with the findings of others, including Robert Putnam in his widely read, very influential book *Bowling Alone* (Putnam, 2000).

In itself, one might respond to this with something like "so what." However, the point is that while Berger seems always to have been basically conservative, his attachment to the interests and aspirations of the American political right, until fairly recently, has not been so starkly in evidence. That, coupled with the strength of his commitment to capitalism, may be unsettling to readers of his work.

Nevertheless, it is very much to his credit that Berger eventually concludes that macro-level organizations – say the NRA or PETA – can also serve as mediating structures. Whatever the strength and character of Berger's commitment to a conservative point of view, it is good to see that he has not wholly and in every instance sacrificed his sociological relevance structure to a naïve American right-of-center ideology.

The fact that modernization and modernity have not lived up to their economic promise is not only cause for discontent, it's a painful fact of everyday life that is routinely denied by politicians and receives too little attention from social scientists and other scholars. Though Berger was not thinking of this when he judged that modernity was an historical concept that, as such, would have an end as well as a beginning, perhaps the end is upon us.

This, I think, is the primary reason why Berger's attention to the youth culture seems excessive. Elsewhere, I dismissed Berger's suggestion that only an affluent society could afford misbehaving children. The disappearance of the youth culture, however, suggests that he may have been right. Participants in the youth culture were secure in the knowledge that their parents – even working class parents – would never let them starve. Now, however, that many working class parents has been reduced to poverty, such assurances are no longer in place.

Furthermore, when good jobs were extraordinarily plentiful, one could drop out for an indefinite period and then re-emerge with the realistic knowledge that he would be able to make a decent living. That time ended during the 1970s. If modernization and modernity gave us the youth culture, it was just on short-term loan. No great loss, perhaps, but surely a failure of modernization and modernity in the nation where they were, and still are, taken most seriously.

Social institutions and distinctively organized activity determine what and how we think and feel, but consciousness reflects back in purposeful and effective ways on the structures that frame it. It's to Berger's credit that he manages to present this scheme simply, plausibly, and dialectically, contributing to his distinctive way of doing the sociology of knowledge, sociological theory, and sociology generally.

Studying and understanding structures and meanings is within the purview of the sociologist, but, as we know, final decisions as to right and wrong remain the responsibility of actors participating in the creation and maintenance of a particular set of social circumstances. Otherwise, sociologists risk becoming ideologues rather than objectively neutral social scientists. Besides, studying and understanding a social context and a normative order are grounded in the empirically accessible world, while judgments as to right and wrong have no such foundation.

The foregoing paragraph gives a dynamic quality to Berger's world view by intimating Weber's apprehension of the ongoing creation of social reality through the accumulation of subjective meanings. The theoretical work, however, is due to Berger, another manifestation of his sociological insight and creativity.

In *The Sacred Canopy* (Berger, 1967: 7) Berger includes a definition similar to this but limited to non-material, man-made culture. This leads him to an interesting conclusion that merits a brief digression peripherally related to our investigation of modernity:

> As but an element of culture, society fully shares in the latter's character as a human product. Society is constituted and maintained by acting human beings. It has no being, no reality, apart from this activity. Its patterns always relative in time and space are not given in nature, nor can they be deduced in any specific manner from the 'nature of man' … one can only say that it is 'the nature of man' to produce a world. What appears at any particular historical moment as 'human nature' is itself a product of 'man's world-building activity' (Berger, 1967: 7).

Society, thus, is an essentially important component of man-made culture. It is structured so as to give order and utility to social life. It allocates resources so as to make the world a livable venue. It assigns people to tasks and positions as demanded by the nature of their cultural context. All this follows from the fact that man is pathetically short on instincts and devoid of an elaborate and finely tuned human nature. As a result, he must create his own world in all its components, and he creates himself in the process (Berger, 1967). Acknowledging this seems a suitable way to close a book on Peter Berger's understanding of modernization and modernity.

Bibliography

Abramsky, S. (2014) *The American Way of Poverty: How the Other Half Still Lives.* New York: Nation Books.

Adorno, T. & Horkheimer, M. (1944/2007) *Dialectic of Enlightenment.* New York: Herder and Herder.

Agrama, H. (2012) *Questioning Secularism.* Chicago: University of Chicago Press.

Althusser, L. & Balibar, E. (1970) *Reading Capital.* Brooklyn, New York: Verso.

American Psychiatric Association (2013) *Diagnostic and Statistical Manual of Mental Disorders: DSM-5.* Arlington, Virginia: American Psychiatric Association.

Amin, S. (2014) *Capitalism in the Age of Globalization.* London: Zed Books.

Andreski, S. (2006) *Max Weber on Capitalism, Bureaucracy, and Religion: A Selection of Texts.* London: Routledge.

Auster, C. (1996) *The Sociology of Work: Concepts and Cases.* Thousand Oaks, California: Pine Forge Press.

Bailyn, B. (2005) *Atlantic History: Concept and Contours.* Cambridge, Massachusetts: Harvard University Press.

Baran, P. & Sweezy, P. (1966) *Monopoly Capital.* New York: Monthly Press.

Becker, R. (1964/1994) *Human Capital: A Theoretical and Empirical Analysis, with Special Reference to Education.* Chicago: University of Chicago Press.

Beeghley, L. (2007) *Structure of Social Stratification in the United States.* Oxford: Routledge.

Bell, D. (1976) *The Cultural Contradictions of Capitalism.* New York: Barnes & Noble.

Bell, D. (1972) *The Coming of Post-Industrial Society.* New York: Basic Books.

Bellah, R., Madsen, R., Sullivan, W., Swidler, A. & Tipton, S. (2007) *Habits of the Heart: Individualism and Commitment in American Life.* Berkeley, California: University of California Press.

Berger, P. (2014) *The Many Altars of Modernity.* Boston, Massachusetts: De Gruyter.

Berger, P. (2011) *Adventures of an Accidental Sociologist.* Amherst, New York: Prometheus Books.

Berger, P. (Ed.) (2009) *Between Relativism and Fundamentalism.* Cambridge, England: William B. Erdmans.

Berger, P. (1992) *A Far Glory.* New York: Anchor.

Berger, P. (Ed.) (1990) *The Capitalist Spirit.* San Francisco, California: Institute for Contemporary Studies.

Berger, P. (1986) *The Capitalist Revolution.* New York: Basic Books.

Berger, P. (1967/1990) *The Sacred Canopy.* New York: Anchor.

Berger, P. (1963) *Invitation to Sociology.* New York: Anchor.
Berger, P. & Huntington, S. (2002) *Globalization.* New York: Oxford.
Berger, P. & Kellner, H. (1981) *Sociology Reinterpreted.* New York: Doubleday.
Berger, P. & Luckmann, T. (1966) *The Social Construction of Reality.* New York: Anchor Books.
Berger, P. & Zijderveld, A. (2009) *In Praise of Doubt.* New York: HarperOne.
Berger, P., Berger, B. & Kellner, H. (1973) *The Homeless Mind.* New York: Vintage Books.
Berger, P., Neuhaus, R. & Novak, M. (1994) *To Empower People: From State to Civil Society.* Washington, D.C.: American Enterprise Institute.
Bickel, R. (2013) *Classical Social Theory in Use.* Charlotte, North Carolina: Information Age.
Bickel, R. & Dufrene, R. (2001) Opportunity and Crime on School Property: A Quantitative Case Study. *Education Foundations.* 15: 19–39.
Bickel, R. & Howley, C. (2003) Math Achievement and the Development of Poor Rural Areas: Effects of Contextual Factors Intrinsic to the Modern World. *Educational Foundations.* 17: 83–105.
Bickel, R. & Howley, C. (2002) Smaller Districts: Closing the Gap for Poor Kids. *American School Board Journal.* 189: 28–30.
Bickel, R. & Lange, L. (1995) Opportunities, Cost, and High School Completion in West Virginia. *The Journal of Educational Research.* 88: 363–370.
Bickel, R. & McDonough, M. (1997) Opportunity, Community and Reckless Lives: Social Distress Among Adolescents in West Virginia. *The Journal of Social Distress and the Homeless.* 6: 29–44.
Bickel, R., Howley, C. & Maynard, A. (2003) No Child Left Behind in Poor Appalachian Districts: Confronting Contextual Factors in the Modern World. *Journal of Appalachian Studies.* 9: 321–340.
Bickel, R., McDonough, M. & Williams, T. (1999) Opportunity, Community, and Early Adolescent Pregnancy: A Replication of Research with Older Teen-Aged Girls. *Educational Foundations.* 13: 47–72.
Bickel, R., Smith, C. & Eagle, T. (2002) Poor Rural Neighborhoods, and Early School Achievement. *Education Policy Analysis Archives.* 6: 9–108.
Bickel, R., Tomasek, T. & Eagle, T. (2000) Top-Down, Routinized Reform in Low-Income, Rural Schools. *The Journal of Poverty.* 8: 1–22.
Bickel, R., Weaver, S., Lange, L. & Williams, T. (1997) Opportunity, Community, and Teen Pregnancy in an Appalachian State. *The Journal of Educational Research.* 90: 175–181.
Blackshaw, T. (2010) *Key Concepts in Community Studies.* Washington, D.C.: Sage.
Blalock, H. (1981) *Social Statistics.* New York: McGraw Hill.
Blau, P. (Ed.), American Sociological Association (1975) *Approaches to the Study of Social Structure.* New York: Free Press.
Blau, P. (1964) *Exchange and Power in Social Life.* Chicago: University of Chicago Press.
Blau, P. (1955) *The Dynamics of Bureaucracy.* Hoboken, New Jersey: Wiley.
Blau, P. & Duncan, O. (1967) *The American Occupational Structure.* Hoboken, New Jersey: Wiley.
Blum, H. (2002) *The Brigade.* New York: HarperCollins.
Bonner, W. & Rajiva, L. (2009) *Mobs, Messiahs, and Markets: Serving the Spectacle in Finance and Economics.* Hoboken, New Jersey: Wiley.

Bottomore, T. (1956) *Selected Writings in Sociology and Social Philosophy.* New York: McGraw-Hill.

Bowles, S. & Gintis, H. (1976) *Schooling in Capitalist America.* New York: Basic Books.

Bowles, S. & Nelson, V. (1974) The "Inheritance of IQ" and the Intergeneration Transmission of Inequality. *The Review of Economics and Statistics.* 56: 39–51.

Bowles, S., Edwards, S. & Roosevelt, F. (2005) *Understanding Capitalism.* New York: Oxford University Press.

Brazinsky, G. (2009) *Nation Building in South Korea: Koreans, Americans, and the Making of a Democracy.* Chapel Hill, North Carolina: UNC Press.

Brown, R. & Vincent, D. (Eds.) (2015) *Social Mobility.* Florence, Kentucky: Routledge.

Burawoy, M. (1992) *Manufacturing Consent.* Chicago: University of Chicago Press.

Burnham, J. (1941/1972) *The Managerial Revolution.* Santa Barbara, California: Praeger.

Callahan, R. (1960) *Education and the Cult of Efficiency.* Chicago: University of Chicago Press.

Charlton, B. (2003) *The Modernization Imperative.* Devon, U.K.: Imprint Academic.

Conrad, J. (1899/2014) *The Heart of Darkness.* Amazon.com: CreateSpace Independent Publishing Platform.

Coser, L. (Ed.) (1975) *The Idea of Social Structure: Papers in Honor of Robert K Merton.* New York: Barnes & Noble.

Cox, O. (1948) *Caste, Class, and Race.* New York: Monthly Review Press.

Crary, J. (2014) *Late Capitalism and the Ends of Sleep.* New York: Verso.

Crooks, R. & Baur, K. (2013) *Our Sexuality.* Independence, Kentucky: Cengage Learning.

Danner, M. (2004) *Torture and Truth: America, Abu Ghraib, and the War on Terror.* New York: New York Review of Books.

Davey, J. (2012) *The Shrinking American Middle Class.* London: Palgrave MacMillan.

Deleuze, G. & Guattari, F. (2009) *Anti-Oedipus: Capitalism and Schizophrenia.* London: Penguin.

Deleuze, G. & Tomlinson, H. (1983) *Nietzsche and Philosophy.* New York: Columbia Classics in Philosophy.

Domhoff, R. (2013) *Who Rules America: The Triumph of the Corporate Rich.* New York: McGraw Hill.

Draut, T. (2007) *Strapped.* New York: Knopf/Doubleday Publishing Group.

Durkheim, É. (1912/1995) *The Elementary Forms of Religious Life.* New York: Free Press.

Durkheim, É. (1898/1986) *Introduction to 4 Major Works.* Beverly Hills, California: Sage.

Durkheim, É. (1897/1997) *Suicide.* New York: Free Press.

Durkheim, É. (1895/2014) *The Rules of Sociological Method.* New York: Free Press.

Durkheim, É. (1893/2014) *The Division of Labor in Society.* New York: Free Press.

Durkheim, É. (1883/1997) *Durkheim's Philosophy Lectures: Notes from the Lycee De Sens Course.* Cambridge, England: Cambridge University Press.

Eagleton, T. (2012) *Why Marx Was Right.* New Haven, Connecticut: Yale University Press.

Eagleton, T. (1976) *Criticism and Ideology.* New York: New Left Books.

Engels, F. (1845/2009) *The Condition of the Working Class in England.* New York: Oxford University Press.

Enteman, W. (1993) *Managerialism: The Emergence of a New Ideology.* Englewood Cliffs, Madison, Wisconsin: University of Wisconsin Press.

Etzioni, A. (1964) *Modern Organizations.* Englewood Cliffs, New Jersey: Prentice-Hall.

Fields, G. (2001) *Distribution and Development: A New Look at the Developing World.* Cambridge, Massachusetts: MIT Press.

Fine, G. & Shulman, G. (2002) *Talking Sociology.* London: Pearson Publishing.

Foucault, M. (1975) *The Archaeology of Knowledge and The Discourse on Language.* New York: Vintage.

Fournier, M. (2012) *Émile Durkheim: A Biography.* New York: Polity.

Freud, S. (1909/1995) *Five Lectures on Psycho-Analysis.* New York: Penguin.

Friedman, M. (2012) *Re-Thinking Anti-Americanism: The History of an Exceptional Concept in American Foreign Relations.* New York: Cambridge University Press.

Friese, S. (2012) *Qualitative Data Analysis with ATLAS.* Washington, D.C.: Sage.

Fromm, E. (1981) *Marx's Concept of Man.* New York: Frederick Unger Publishing.

Gadamer, H-G. (1960/2004) *Truth and Method.* New York: Crossroad.

Garfinkel, H. (1967) *Studies in Ethnomethodology.* New York: Polity.

Gaskell, P. (1833/2010) *The Manufacturing Population of England: Its Moral, Social, and Physical Conditions, and the Changes Which Have Arisen from the Use of Steam Machinery; With an Examination of Infant Labour.* Whitefish, Montana: Kessinger Publishing.

Gerth, H. & Mills, C. (1958) *From Max Weber.* New York: Oxford University Press.

Glazer, N. (Ed.) (1992) *From Kafka: The Complete Stories and Parables.* New York: Quality Paperback Book Club.

Glassman, R., Swatos, W. & Rosen, R. (1987) *Bureaucracy and Democracy and Socialism.* New York: Greenwood Press.

Goodsell, C. (2014) *The New Case for Bureaucracy.* Washington, D.C.: Sage.

Goodsell, C. (1983) *The Case for Bureaucracy: A Public Administration Polemic.* Chatham, New Jersey: Chatham House Publishers.

Gormley, W. & Balla, S. (2012) *Bureaucracy and Democracy: Accountability and Performance*, 3rd Edition. Washington, D.C.: CQ Press.

Gouldner, A. (1980) *The Coming Crisis of Wester Sociology.* New York: Basic Books.

Gross, M. (1977) *The World of Raymond Chandler.* New York: A&W.

Habermas, J. (2015) *The Lure of Technocracy.* New York: Polity.

Habermas, J. (1975) *Legitimation Crisis.* Boston, Massachusetts: Beacon Press.

Harvey, D. (2010) *A Companion to Marx's Capital.* New York: Verso.

Hayek, F. (1944/2007) *The Road to Serfdom.* Chicago: University of Chicago Press.

Harrison, L. & Huntington, S. (2001) *Culture Matters.* New York: Basic Books.

Henderson, A. & Parsons, T. (1947) *Max Weber: The Theory of Social and Economic Organization.* Eastford, Connecticut: Martino Fine Books.

Hickling-Hudson, A. & Klees, S. (2012) Alternatives to the World Bank's Strategy for Education and Development in Klees, S., Samhoff, S. & Stromkoes, N. (Eds.) *The World Bank and Education: Critiques and Alternatives.* Rotterdam, Netherlands: Sense Publications, 205–226.

Himmelfarb, G. (1984) *The Idea of Poverty.* New York: Vintage.

Hoffer, E. (2006) *Reflections on the Human Condition.* Titusville, New Jersey: Hopewell Publications.

Horkheimer, M. (1947/2013) *Eclipse of Reason.* Eastford, Connecticut: Martino Fine Books.

Horkheimer, M. & Adorno, T. (1947/2007) *Dialectic of Enlightenment*. Stanford, California: Stanford University Press.

Howley, A., Bickel, R. & McDonough, M. (1997) The Call for Parent Involvement in Rural Communities: Mantra and Mystification. *The Journal of Research in Rural Education*. 13: 101–107.

Hox, J. & Roberts, J. (2010) *Handbook of Advanced Multilevel Analysis*. Florence, Kentucky: Routledge.

Hubbard, K. (2013) *Serving Victoria*. New York: Harper Perennial.

Hugo, V. (1862/2013) *Les Misérables*. New York: Berkeley Publishing Group.

Iceland, J. (2003) *Poverty in America: A Handbook*. Berkeley, California: University of California Press.

Jessop, B. (1987) *The Economy, The State, and The Law*. Florence, Italy: European University Institute.

Kahn, H. (1979) *World Economic Development: 1979 and Beyond*. New York: William Morrow.

Klein, N. (2007) *The Shock Doctrine*. New York: Henry Holt Books.

Krasko, G. (2004) *The Unbearable Boredom of Being*. Milford, Connecticut: Universe Publishing.

Krugman, P. (2009) How Did the Economists Get It So Wrong? *New York Times Magazine*. 2 September: 2–13.

Kuhn, T. (1962) *The Structure of Scientific Revolutions*. Chicago: University of Chicago Press.

Kuznets, S. (1968) *Toward a Theory of Economic Growth*. New York: W.W. Norton and Company.

Liebow, L. (1995) *Tell Them Who I Am*. London: Penguin Books.

Lipset, S. (1960) *Political Man: The Social Bases of Politics*. New York: Doubleday.

Lockhart, P. (2014) *Measurement*. Belknap, Illinois: Belknap Publishers.

McClay, W. (2009) *Mediating Institutions*. Washington, D.C.: American Enterprise Institute.

McGuire, M. (2008) *Religion: The Social Context*. Long Grove, Illinois: Waveland Press, Inc.

McShane, P. (2016) *Profit: The Stupid View of President Donald Trump*. Vancouver, Canada: Axial Publishing.

Maguire, E. (2003) *Organization Structure in American Police Agencies*. Albany, New York: State University of New York Press.

Mandel, E. (1994) *The Place of Marxism in History*. New York: Prometheus Books.

March, J. & Simon, H. (1958) *Organizations*. Hoboken, New Jersey: Wiley.

Marcuse, H. (1964) *One Dimensional Man*. Boston, Massachusetts: Beacon.

Martin, J. (2008) *The Poulantzas Reader*. London: Verso.

Marx, K. (1867/2011) *Capital: Volume 1*. London: Penguin.

Marx, K. (1844/1988) *The Economic and Philosophic Manuscripts of 1844*. Amherst, New York: Prometheus Books.

Marx, K. & Engels, F. (1848/2014) *The Communist Manifesto*. New York: International Publishing Company.

Marx, K. & Engels, F. (1846/1998) *The German Ideology*. Amherst, New York: Prometheus Books.

Mead, G. (1934/1967) *Mind, Self, and Society: From the Standpoint of a Social Behaviorist*. Chicago: University of Chicago Press.

Merton, R. (1949/1968) *Social Theory and Social Structure.* New York: The Free Press.

Mills, C. (1959) *The Sociological Imagination.* New York: Oxford University Press.

Moynihan, D. (1965) *The Negro Family: The Case for National Action.* United States: Office of Planning and Research. United States Department of Labor.

Neuendorf, K. (2001) *The Content Analysis Guidebook.* Washington, D.C.: Sage.

Newman, K. (1999) *Falling from Grace: Downward Mobility in the Age of Affluence.* Berkeley, California: University of California Press.

Newman, K. (1993) *Declining Fortunes: The Withering of the American Dream.* New York: Basic Books.

Noble, D. (1984) *Forces of Production.* New York: Knopf.

Noble, D. (1979) *America by Design.* New York: Oxford University Press.

Ollman, B. (1977) *Alienation: Marx's Conception of Man in Capitalist Society.* Cambridge, England: Cambridge University Press.

Parsons, T. (1971) *Societies: Evolutionary and Comparative Perspectives.* Englewood Cliffs, New Jersey: Prentice Hall.

Parsons, T. (1951/2012) *The Social System.* New Orleans, Louisiana: Quid Pro Books.

Parsons, T. (1937/1967) *The Structure of Social Action.* New York: The Free Press.

Paxton, P., Hipp, J. & Marquart-Pyatt, S. (2011) *Nonrecursive Models.* Washington, D.C.: Sage.

Perkinson, H. (1968/1995) *The Imperfect Panacea.* New York: Random House.

Perrow, C. (2014) *Complex Organizations: A Critical Essay.* Brattleboro, Vermont: Echo Point Books and Media.

Pfadenhauer, M. (2013) *The New Sociology of Knowledge.* London: Transaction.

Piketty, T. (2014) *Capital in the Twenty-First Century.* Cambridge, Massachusetts: Harvard University Press.

Poston, D. & Bouvier, L. (2010) *Population and Society.* Cambridge, England: Cambridge University Press.

Poulantzas, N. (1978/2015) *State, Power, Socialism.* New York: Verso.

Putnam, R. (2015) *Our Kids: The American Dream in Crisis.* New York: Simon & Schuster.

Putnam, R. (2000) *Bowling Alone: The Collapse and Revival of American Community.* New York: Simon & Schuster.

Ravitch, D. (2014) *Reign of Errors.* New York: Vintage.

Ritzer, G. (2004) *The McDonaldization of Society.* Thousand Oaks, California: Pine Forge.

Robertson, G. (2006) *Crimes Against Humanity: The Struggle for Global Justice.* London: Penguin Books.

Robinson, J. (1942/1967) *An Essay on Marxian Economics.* London: Palgrave MacMillan.

Robinson, N. (2013) *Economic Principles and Issues.* Dubuque, Iowa: Kendall Hunt Publishing.

Ropper, A. & Samuels, M. (2009) *Adams and Victor's Principles of Neurology.* New York: McGraw Hill Professional.

Roth, G. & Wittich, C. (Eds.) (1978) *Max Weber: Economy and Society.* Berkeley, California: University of California Press.

Santayana, G. (1967) *Animal Faith and Spiritual Life.* New York: Appleton-Century-Crofts.

Sartre, J-P. (1943/1993) *Being and Nothingness.* New York: Washington Square Press.

Schafft, K. & Jackson, A. (2010) *Rural Education for the Twenty-First Century.* University Park, Pennsylvania: Penn State University Press.

Schumpeter, J. (1942/2008) *Capitalism, Socialism, and Democracy.* New York: Harper.

Schutz, A. (1967) *The Phenomenology of the Social World.* Evanston, Illinois: Northwestern University Press.

Schutz, A. & Luckmann, T. (1973) *The Structures of the Life World, Volume 1.* Evanston, Illinois: Northwestern University Press.

Schwarz, J. (1998) *Illusions of Opportunity: The American Dream in Question.* New York: W.W. Norton and Company.

Scott, W. (2013) *Institutions and Organizations.* Washington, D.C.: Sage.

Simmel, G. (1910) *Hauptprobleme der Philosophe.* Leipzig, Germany: Goschen.

Smith, L-P. (Ed.) (1920/2008) *Little Essays: The Writings of George Santayana.* London: Forgotten Books.

Smith, M. & Marx, L. (1994) *Does Technology History: The Dilemma of Technological Determinism.* Cambridge, Massachusetts: Harvard University Press.

Sloterdijk, P. (2013) *The World Interior of Capitalism.* New York: Polity.

Sokal, A. & Bricmont, J. (1999) *Postmodern Intellectuals Abuse of Science.* London: Macmillan.

Sowell, T. (2014) *Intellectuals and Society.* New York: Basic Books.

Stiglitz, J. (2003) *Globalization and Its Discontents.* London: Penguin.

Stroup, G. (2009) *Abingdon Pillars of Theology.* Nashville, Tennessee: Abingdon Press.

Sweezy, P. (1981) *Four Lectures on Marxism.* New York: Monthly Review Press.

Sweezy, P. (1970) *Theory of Capitalist Development*, 2nd Edition. New York: Monthly Review Press.

Tawney, R. (1926/2005) *Religion and the Rise of Capitalism.* Dublin, Ireland: Mentor.

Thompson, E. (1963/1980) *The Making of the English Working Class.* London: Victor Gollancz.

Thurow, L. (1980/2001) *The Zero-Sum Society.* London: Penguin.

Tinbergen, J. (1956) *Economic Policy: Principles and Design.* Amsterdam, Netherlands: North-Holland Publishing.

Todaro, M. (1985) *Economic Development in the Third World*, 3rd Edition. London: Longman Publishing.

Tolman, E. (1948) The Cognitive Maps of Rats and Men. *Psychological Review.* 55: 189–208.

Tönnies, F. (1887/2001) *Gemeinschaft and Gesellschaft.* Cambridge, England: Cambridge University Press.

Trotsky, L. (1936/2004) *The Revolution Betrayed.* Mineola, New York: Dover Books.

Trotsky, L. (1930/2007) *The History of the Russian Revolution.* Chicago: Hay Market Books.

Tu, W. (1979/1996) *Confucian Traditions in East Asian Modernity: Moral Education and Economic Culture in Japan and the Four Mini-Dragons.* Cambridge, Massachusetts: Harvard University Press.

Uchitelle, L. (2007) *The Disposable American: Layoffs and Their Consequences.* New York: Vintage Books.

Updike, J. (1971) *Rabbit Redux.* New York: Alfred A. Knopf.

Veblen, T. (1923/1997) *Absentee Ownership in Recent Times: The Case of America.* New York: B.W. Huebsch.

Veblen, T. (1921/2013) *The Engineers and the Price System.* Memphis, Tennessee: General Books.

Veblen, T. (1918/2015) *The Instinct of Workmanship and the State of the Industrial Arts.* London: Fb &C Limited.

Veblen, T. (1904/2013) *The Theory of Business Enterprise.* New York: Charles Scribner's Sons.

von Mises, F. (1944/1996) *Bureaucracy.* Alexandria, Virginia: Libertarian Press.

Wagner, P. (1994) *A Sociology of Modernity, Liberty, and Dignity.* Abington-On-Thames, U.K.: Routledge.

Walkey, F. & Welch, G. (2010). *Demystifying Factor Analysis.* Victoria, British Columbia, Canada: Xlibris Corporation.

Weber, M. (1922/1978) *Economy and Society: An Outline of Interpretative Sociology.* Berkeley, California: The University of California Press.

Weber, M. (1920/2010) *The Protestant Ethic and the Spirit of Capitalism.* New York: Oxford University Press.

Whitaker, J. (1990) *Centenary Essays on Alfred Marshall.* Cambridge, England: Cambridge University Press.

Wiebe, R. (1966) *The Search for Order, 1877–1920.* New York: Hill and Wang.

Williams, R. (1978) *Marxism and Literature.* New York: Oxford University Press.

Willis, P. (1981) *Learning to Labor.* New York: Columbia University Press.

Wilson, K. (1997) Science and Treatment Development: Lessons from the History of Behavior Therapy. *Behavior Therapy.* 28: 547–558.

Wood, P. (2014) *Technocracy Rising: The Trojan Horse of Global Intelligence.* Mesa, Arizona: Coherent Publishing.

Wright, E. (2015) *Understanding Class.* New York: Verso.

Wright, E. (1979) *Class Structure and Income Determination.* New York: Academic Press.

Zizek, S. (2011) *Living in the End Times.* New York: Verso.

Index

For Product Safety Concerns and Information please contact our EU representative GPSR@taylorandfrancis.com
Taylor & Francis Verlag GmbH, Kaufingerstraße 24, 80331 München, Germany

www.ingramcontent.com/pod-product-compliance
Ingram Content Group UK Ltd.
Pitfield, Milton Keynes, MK11 3LW, UK
UKHW011011180926
475701UK00004B/676

* 9 7 8 0 3 6 7 2 0 8 1 6 5 *